DESMOND MICHAEL COVERLEY, Ph.D.

Climate Change

and

SIGNS

of

The 'End Time'

This book explains how the Events of Climate Change
ARE SIGNS Of the 'End Time'

Publisher:

This book may be ordered through booksellers – Kindle Direct Publishing

or

by contacting:

Corban Jireh Enterprises (CJE)
3308 Clay Street,
Silver Spring, Maryland 20902
Email: cjenterprises7@outlook.com
Website: www.corbanje.com;
Telephone (301) 448-9271

****

The Author's information may be viewed on the last page.
Corban Jireh Enterprises: 2021

****

ISBN: 978-1-64826-493-1
This Edition is Reviewed and Printed by
Kindle Direct Publishing -USA

Contents

Preface

The news media have brought to the world's attention the strange and unusual that are taking place on Planet Earth, in its sky, on its land, in its waters, and with its weather. Each day people throughout the world look for answers for these events. Therefore, this book emerges with the answers.

First, this book's goal is to find and explain the truth of these strange and unusual events around the world and to give real answers. To satisfy this goal, the author took the following actions:

- There was an examination of the stories in the news media.
- The sharing of the feelings of the lay people around the world.
- The author tested the answers given by the scientific community for the strange and unusual events.
- The author unveiled the teachings of the prophetic scriptural texts of the Holy Bible.
- The author compared the predictions of the scientific community with the predictions of the Holy Bible.
- People who read this book will realize that the strange and unusual events of Climate Change are signs of the "End Time."

Second, after the author examined the news stories, the feelings of laypeople, and the theories and predictions of the scientific community, He examined the prophetic scriptural texts of the Holy Bible. The prophecies made by Jesus Christ, his apostles, and many of the Old Testament prophets all have events that are strange and unusual that will occur in the sky, on the land, in the waters, and with the weather. Also, the target of these strange and unusual events are predicted in the Bible to target Planet Earth in *"the last days, end of the world, or end time."* Interestingly, the prophetic events predicted by the Bible, are like those that science presents as events of Global Climate Change. Because of scientific belief, the book carefully examines the "End Time Signs" with the "Global Climate Change Events" to see if they are different or the same.

Third, in carrying out this process of examination between Global Climate Change and Prophetic Predictions of the Bible End Time Events, the author explained the meaning of "Signs" and "Events" as used in the book. So, because a "Sign" is an object, quality, event, or entity whose presence or occurrence indicates the probable presence or occurrence of something else, in this book, it is referred to as "Prophetic Events of the Bible about the "End Time or Last Days." So, "sign' or "event" will have the same meaning. Therefore, when comparing the "events" of the scientific community on global warming/climate change and the "signs" of the prophetic predictions of the Holy Bible, the author seeks to find if the "events" of global climate change are the Bible prophetic "signs" of the end time or last days. Therefore, the investigation into the strange and unusual "events or signs" was critical for many reasons:

- The news media want to tell the real story to people.
- The lay-people of the world want the truth about the events because many such events are troublesome and scary.
- All people do not accept the scientific community's predictions. They feel it needs more credibility.
- The need to review the findings of science considering the prophecies of the Holy Bible.
- The people who lean toward the teachings of the prophecies of the Holy Bible want to know if the events that are taking place are true "Signs of the End Time."

Fourth, the book presents a careful comparison of science presentations and predictions of global warming/climate change and the predictions of the prophetic scriptural texts. The author examined the prophetic predictions of the *'End Times'* events of the Old Testament prophets, Jesus Christ, and the New Testament apostles. The findings predict strange and unusual events/signs in the sky, on the land, in the waters, and with the weather.

Fifth, concerning signs in the sky and with the weather, (1) the sign of strange and unusual death of birds that fly in the sky. (2) Unusual torrents of rain, snow, hail and hailstones, lightning, and thunderstorms, and even fire

coming from the sky to the earth. (3) The heavenly bodies will present strange and unusual signs. For example, the sun, moon, and stars will not give any light to the earth. Therefore, the earth will be in darkness. Then again, the sun will become so hot that it will burn the skins of the people of earth.

Sixth, concerning 'Signs' on the Land and in the waters, the Bible predicts that on the land there will be strange and unusual famines, pestilences, and wars among the people of the earth. Also, there will be great hate among the people for each other and also for God. Signs in the waters will also occur. Drinking water will not be available and the oceans' roaring will have a great effect on the people. Really, there are so many more these prophetic scriptures God will present with many unusual signs. However, all such eventful signs of the *Last Days* or *End Time* are presented in this book.

Seventh, the belief is that these strange and unusual events that are taking place in the sky, on land, in the waters, and with the weather are God's signs that time is short. It is interesting, however, that the scientific community is predicting what the Bible has prophesied. Therefore, the author wants people to read this book, examine the findings between Global Climate Change and the Prophetic Signs of the End Time. Then, see if these strange and unusual events are because of global warming with its resulting Climate Change or, "Are they Signs of Warning from God?" Could it be that Climate Change, which results from Global Warming, is an instrument that God is using to warn people about the *End Time*?

Eighth, this book explains and presents the truth about the strange and unusual signs by linking the *stories of the news media*, the *feelings of the lay-people of the world*, the *predictions of the scientific community*, and the *biblical prophetic predictions about End Times' Signs*. The book presents this important information in *Six Sections and*

23 short chapters. Section One presents *Strange and Unusual Signs Occurring on Planet Earth.* Section Two presents *The Science of Global Warming/Climate Change.* Section Three presents *The Bible Old Testament Prophets Predict Unusual 'End Time' Signs.* Section Four presents *The Bible New Testament Gospels Predict Unusual 'End Time' Coming Signs.* Section Five presents *The Bible New Testament Epistles Predict Unusual 'End Time' Coming Signs.* Section Six presents *Decisions.* Get one, read it, and share the information with your family, friends, churches, & social groups.

Tribute And Acknowledgments

First, the author wants to thank God for His love for all of us and for providing eternal life for all people, irrespective of race, creed, skin color, or ethnic backgrounds. The fact is, thanks to His Word, the Bible, which reminds us:

> For so loved God the world that the Son He gave, **The Only Begotten One**, that **all those 'who are believing'** in Him will not perish but have eternal life. God did not send his Son into the world to condemn it, **but to save all the people of the world** through His Son. **Any person** who believes in Him, the Son, God does not condemn (John 3:16- 19).
>
> And they sang a new song, saying, 'Thou art worthy to take the book, and to open the seals thereof: for you died, and you have redeemed us to God by thy blood out of **every kindred**, and **tongue,** and **people,** and **nation** (see Revelation 5:9).

We thank and praise God for His concern for all people. In the agreement of the Godhead to save all of us, The Godhead agreed that The Son, The Only Begotten One, a Person of the Godhead would come to earth, suffer, and die on the cross of Calvary for us all. Also, all people who are believing in Him, The Only Begotten One, He will give to them eternal life.

Second, the author wants to acknowledge and give a Tribute to the late **Dr. Gene Gary Williams, former Dean of the College of Allied Health Sciences at Howard University, Washington, DC.** He thanks her for editing his book: "The Bible Explains the Climate Change Controversy." This was her very last accomplishment. The same day she completed the editing of the book, she went to be with the LORD in heaven. We will ever remember Dr. Gene Gary Williams, Ph.D. for her contributions to us while she was on the earth. May God bless her family, especially Gina and her family, who are left behind. Get the book from Amazon, under, Books by Desmond Michael Coverley, Ph.D.

Introduction

THE SIGNIFICANCE OF THIS BOOK

First, this book is significant because of the promises it shares from God's Word:

> This is what the Sovereign LORD asks: Are you the one I was talking about long ago, when I announced through Israel's prophets that in the future I would bring you against my people? But this is what the Sovereign LORD says: When Gog invades the land of Israel, my fury will boil over! In my jealousy and blazing anger, I promise a mighty shaking in the land of Israel on that day. All living things—the fish in the sea, the birds of the sky, the animals of the field, the small animals that scurry along the ground, and all the people on earth—will quake in terror at my presence. Mountains will be thrown down; cliffs will crumble; walls will fall to the earth. I will summon the sword against you on all the hills of Israel, says the Sovereign LORD. Your men will turn their swords against each other. I will punish you and your armies with disease and bloodshed; I will send torrential rain, hailstones, fire, and burning sulfur! In this way, I will show my greatness and holiness, and I will make myself known to all the nations of the world. Then they will know that I am the LORD (EZEKIEL 38:17-23).

> What sign will signal your return and the end of the world?" Jesus told them, "Don't let anyone mislead you, for many will come in my name, claiming, 'I am the Messiah.' They will deceive many. And you will hear of wars and threats of wars, but don't panic. Yes, these things must take place, but the end won't follow immediately. Nation will go to war against nation, and kingdom against kingdom. There will be famines and earthquakes in many parts of the world. But all this is only the first of the birth pains, with more to come "Then you will be arrested, persecuted, and killed. You will be hated all over the world because you are my followers. (Matthew 24:3-9).

Second, this book is significant because of the Summation of its Researched Materials.

(1) This book explains how the events of Global Climate Change are SIGNS OF THE 'END TIME'. In presenting this evidence, first, the book tells how 'Strange and Unusual Signs' are occurring on Planet Earth. The signs are evident in the Sky, on the Land, in the Waters, and with the Weather. In each of the mentioned areas, the lives of animals and people are affected by sickness and death.

(2) The book presents scientific evidence and tells how science makes a claim that the signs in the sky, on the land, in the waters, and with the weather are because of Global Climate Change. Besides, science states that the cause of the events is the warming of the earth by people. Also, if this warming process does not stop, Planet Earth will end.

(3) The book does not deny science's fact that Global Climate Change is real, but it shows that the evident SIGNS presented are of a different origin and that climate change is a tool. So, the events that science is calling events of Global Climate Change are SIGNS OF THE 'END TIME' which are the exact predictions of the Bible.

(4) The book then presents and explains from the prophecies of the Old Testament prophets, the Gospels of Jesus Christ, and Epistles of the New Testament, the proof of occurring signs.

(5) The proof is overwhelming, to see how God has endowed the scientists to make predictions about Global Climate Change, which is online with the predictions of the Bible, God's Holy Word.

(6) The book concludes by giving the readers an opportunity TO THINK ABOUT a few things: (a) The Truth of the Events of Climate Change and the Signs of the 'End Time', (b) Why is God Using Strange and Unusual Signs? and (c), What is Your Decision? Readers will not be disappointed in their reading of this book. Get a copy from Amazon, read, see, and learn how God is still in charge of all the events that are occurring on Planet Earth.

SECTION ONE

Strange and Unusual 'Signs' are Occurring on Planet Earth

Strange and unusual stories are unfolding by the News Media of strange and unusual signs in the sky, on land, in the waters, and with the weather. These stories are so strange and unusual and there are so many, even this book cannot share them all. The problem is, as soon as the news media reveal an unusual story, there are others taking place. The other problem, according to the news media, science and laypeople do not have any real answers for these signs that are taking place over the entire world. However, someone needs to give answers. After all, people need to know what is going on in the world. Since science does not give the answers many people are seeking, many of them are turning to their religious books, documents, teachers, preachers, ministers, and other means for interpretation. Many people believe that we are living in the *'End Times'* and God is trying to get our attention through the manifestation of the strange and unusual signs that are occurring. Therefore, we must investigate the stories that are being shared by the news media, the information presented by science, and the prophecies of the Bible, so we can make our conclusions about the strange and unusual signs which are occurring, **in the sky, ~⸎~ on the land, ~⸎~ in the waters, ~⸎~ and with the weather.**

Chapter 1

STRANGE AND UNUSUAL 'SIGNS' ARE OCCURRING IN THE SKY

The news media presented stories of hundreds, even thousands of birds falling from the sky dead or dying. The press labeled many of the strange and unusual deaths as mysterious. Also, the deaths of the birds are baffling to scientists. Of the thousands and thousands of stories about massive bird deaths, this chapter presents news media stories of mass bird deaths from 2011 to 2019.

Sky Sign No. 1: Birds falling from the sky

Photo of birds falling from the sky in Australia - dead

The news media presented stories of hundreds, even thousands of birds falling from the sky dead or dying. The press labeled many of the strange and unusual deaths as mysterious. Also, the deaths of the birds are baffling to scientists. Of the thousands and thousands of stories about massive bird deaths, some are presented in this chapter. News media stories of mass bird deaths from 2011 to 2019 are given.

The Year 2011 Mass Bird Deaths

In 2011 birds died all over the world. *First,* on December 20, people found four hundred birds dead in Davis County, Utah. The news revealed that officials were "unsure of the cause of the mass death." In one case they referred to the deaths as "mysterious" (Morgan 2011; the *Associated Press* 2011). *Second,* on July 15, 2011, hundreds of thousands of prions (a type of seabird) people found dead in New Zealand. Officials said there was no known cause for the death of these birds, but they blamed the weather as the cause of their deaths (Graham 2011). *Third,* on August 8, 2011, eight thousand turtle doves fell from the sky, dead, in Faenza, Italy. The people expressed much frustration over this unusual event because the authorities did not know the cause (Longbottom 2011).

The Year 2012 Mass Bird Deaths

The year 2012 witnessed some of the same stories, where birds fell dead from the skies. For example, 465 events took place in sixty-seven countries. Among these mass deaths were fowls of the air, the land, and the water (End Time Prophecy 2014). Again, this chapter shares a few cases with the readers.

First, the November 20, 2012, incident that happened on a road in Missouri, people discovered a flock of birds that had fallen from the sky, dead. This event again baffled the scientists (Fox2Now, St. Louis 2012). *Second,* on November 12, 2012, in England, thousands of birds dropped out of the sky into the sea to their deaths. Species included goldcrests, robins, thrushes, and blackbirds. The sky was thick with garden birds. As they fell to the ground, witnesses said the birds appeared exhausted and disoriented. As a result, the thousands that fell in areas that had water, drowned, stated eyewitnesses (Godfrey 2012). *Third,* another strange and unusual event occurred in Chile on May 12, 2012. The people there stated that 2,300 birds died on their beaches, according to Marilia Brocchetto of CNN.

These are just a few of thousands and thousands of cases of the deaths of birds in 2012 around the world. Similar stories, however, continued into 2013 and up through 2016.

The Year 2013 Mass Bird Deaths

Records of bird deaths in 2013 showed that 798 known mass death events occurred in ninety-three countries (see Appendix No. 2) to view the online statistics). Again, the chapter shares a few of these cases with the readers from the many incidents that took place around the world.

First, on December 6, 2013, thousands of water birds died at Kamfers Dam in South Africa. The death of these birds puzzled people, including the authorities (Wildenboer 2013). *Second,* on November 26, 2013, thousands of dead birds washed ashore on Saint Lawrence Island, Alaska. The authorities stated that the death of the birds was because of the frigid weather (DeMarban 2013). *Third,* on July 19, 2013, in Queensland, Australia, several black kite birds fell from the sky, dead. The headline of the *Courier Mail* read, "Bird experts and scientists left puzzled as birds fall dead from north Queensland skies." It reported that experts were looking for clues on why this occurred (Michael, 2013). *Fourth*, on April 17, 2013, thousands of birds dropped dead from the sky at an army base in Utah. The authorities had no sound answers for this occurrence when challenged by the news media (*Associated Press*). *Fifth,* on April 8, 2013, thousands of seabirds washed up dead on the northeast coast of England. Here, like many of the others, they blamed the weather (Wood 2013) (Appendix No. 3 online statistics should be of interest).

The Year 2014 Mass Bird Deaths

The year 2014 did not differ from previous years. End Time Prophecy (2014) stated that 651 known events of massive animal deaths occurred in seventy-six countries. The chapter identifies five cases to share with readers.

First, on November 23, 2014, the news media reported the mass deaths of seabirds. These birds washed up along the Sonoma Coast in

California (Callahan 2014). *Second*, on September 22, 2014, hundreds of dead birds washed ashore because of weather in East Lothian, Scotland (BBC NEWS 2014). *Third,* on July 4, 2014, the citizens found hundreds of dead birds on a beach after a storm at Lake Winnipeg in Canada (CTV News– Winnipeg 2014). *Fourth,* on June 19, 2014, when concerned citizens found seabirds dead along the coast of Port Elizabeth, South Africa, the authorities called it "mysterious and worrying." Answers were inconclusive (Williams 2014). *Fifth,* on June 26, 2014, the people of Peru found 1,832 dead seabirds on the coast of Arequipa, Peru. The reason given was that it was "apparently because of the displacement of warm waters to the south because of El Niño" (Del Perú Para Elmundo 2014). Again, the answer to this incident was the *change in weather* (see Appendix No. 4 to view the online statistics for additional stories of bird deaths).

The Year 2015 Mass Bird Deaths

The event summary for bird deaths in 2015 is, 544 known mass death events in eighty countries, as of August 2, 2015 (see appendices - End Time Prophecies 2015).

First, on January 7, 2015, CBC News, stated that the people of British Columbia since October 2014 found over 100,000 dead seabirds along the west coast of America.

Second, on May 18, 2015, in Hualpen, Chile, the citizens found 1,300 seabirds dead on the beaches. Also, other dead birds were found at Ramuntcho Lenga Bay in Chile, the reason unknown. Officials of the Agriculture and Livestock Service (SAG) took samples of the birds to look for clarification for this strange and unusual event. They wanted to know if disease, pollution, or some action by sardine fishermen caused the event (Tooley 2015). *Third,* hundreds of birds died in El Reno, Oklahoma. The remains of dead birds were in the trees, on telephone wires, and the ground. A truck driver stated that he'd lived in Oklahoma all his life and never had seen anything like the deaths of the birds. Again, the cause of the bird deaths remains a mystery (Passoth 2015) (see Appendix No. 5 to view the online statistics).

The Year 2016 Mass Bird Deaths

In 2016, the strange events of bird deaths continued. *First,* on January 12, 2016, the *Washington Post* ran a story about the mysterious deaths of birds in Alaska, noting, "Tens of thousands of dead birds are washing up on the beaches of Alaska's Prince William Sound." Some experts say the changing climate caused the unexplained mass die-off of the birds (Warrick 2016) (see Appendix No. 6 to view the online statistics of the masses of bird death for the year 2016).

The Year 2017 Mass Bird Deaths

According to *End Time Prophecies,* the summary for 2017 of mass animal deaths events in approximately eighty-two countries, some 454 known mass deaths occurred. Because of the many cases of bird mass deaths, the chapter refers the reader to the Appendices for the individual cases for 2017 to 2019. Readers will find that the mass deaths are increasing (see Appendix No. 7 to view the online statistics of bird deaths).

The Year 2018 Mass Bird Deaths

The summary for 2018 of mass animal deaths events in approximately sixty-seven countries, some 388 known mass deaths occurred (see Appendix No. 8 to view the online statistics for these unusual cases).

The Year 2019 Mass Bird Deaths

Besides all the mass deaths of birds falling from the skies, for 2019, so far, the records show, thousands of birds are mysteriously dying. For example, to view the entire cases of mass deaths, see Appendix No. 9. However, as of September 19, 2019, the statistics show for America alone, "3 Billion birds have died off during the past 50 years in America." (Link)

Scientists Struggle for Answers

Along with the birds falling from the sky, the statistics show that thousands of chickens, ducks, pelicans, and other fowls died from various diseases, especially avian flu. On March 4, 2015, 4.17 million birds had died since January 2015 due to avian flu in Taiwan, China (Focus Taiwan 2015). These millions of bird deaths took place in that nation alone. There appears to be no clear explanation for the unusual deaths of land birds and seabirds, but the scientists gave two suggestions. One was, *the drastic shift in ocean temperature*, and the other, *the change in the temperature is affecting the feeding conditions for the birds*. However, many of the news media – The Press Democrat of California, Scotland news, and the CTV News of Winnipeg, Canada, all concluded that climate change influences the birds' deaths (Welch 2015).

The above is about the first strange and unusual events of birds falling from the sky to the grounds of the earth, dead. The second strange and unusual sign or event is about meteorites falling from the sky. Let us investigate some of these events.

Sky Sign No. 2: Meteorites Falling from the Sky

Photo showing a falling Meteorite

A strange phenomenon in the universe is that meteorites appear to become closer and closer to striking the earth. The *Associated Press* and the *Global News* released frightening information on February 15, 2013, concerning a meteorite that smashed into Russia. The event left several interesting areas of concern. ABC News (2013) stated that scientists estimated that the object weighed about ten tons and was moving at a speed of about 33,554 miles per hour as it approached the earth's atmosphere. Also, upon impact, it set off a sonic blast that injured over one thousand people, damaged hundreds of buildings, and created panic across the region (ABC News, February 19, 2013). Information received from NASA showed that the meteor released about 300 kilotons of energy—the largest in over a century—which is about twenty-five times more potent than the atomic bombs dropped in World War II (NASA).

The Russian prime minister, Dmitry Medvedev, stated that the meteor showed the vulnerability of Planet Earth. Also, Deputy Prime Minister Dmitry Rogozin stated that the world is unsafe because neither Russia nor the United States has the technology to shoot down meteors. An American national security reporter Spencer Ackerman agreed that no known weaponry on earth can prevent meteors from crashing into the ground and that we are "sitting ducks."

The above meteor held the headlines for a few years. However, according to Brett Molina of USA TODAY, since the 2013 incident over Russia, there was another event that occurred in December of 2018. The meteor, according to the National Aeronautics and Space Administration (NASA), (1) gave off an explosion 15 miles away from earth, over the Bering Sea, (2) contained over 10 times the energy of Hiroshima, and (3), it was the largest recorded event since 2013 incident in Russia (Molina, USA TODAY, 2019).

According to the news media, many laypeople and the clergy believe that meteorites falling to earth are warnings from God. However, in addition to meteorites falling from the sky, there is another sign in the sky. This one is that of cyber warfare.

Sky Sign No. 3: Cyber Warfare Rages Across the Sky

Meaning of Cyber Warfare

Cyberwarfare is "actions by a nation-state to penetrate another nation's computers or networks to cause damage or disruption" (Clarke 2010). Today with the use of computers and the internet, terrorist groups, political or ideological extremist groups, hackers, and criminal organizations are all engaged in cyber warfare.

The news media reminds us that people do not always take hacks and malware seriously, as they relate to private and government computer systems. People feel anti-malware will clean up the computers. But hackers have become more sophisticated at breaking into computers. Their attempts have moved to the level of war, and this "cyberwar" is on individuals, companies, organizations, and government systems of security.

James R. Clapper's View of Cyber Warfare

James Clapper, the United States Director of National Intelligence gives his view on Cyber Warfare. He states that the US has divided cyber- attacks into two broad categories. These are *cyber-espionage* and *cyber- attacks*. *Cyber-attacks* are the top security threat to the United States. *We know cyber espionage*, as the act or practice of getting secrets such as sensitive or classified information from individuals, competitors, rivals, groups, governments, and enemies.

Using this information is for military, political, or economic advantage. To gather such information, hackers use illegal exploitation methods on the Internet, networks, software and or computers (Clapper 2015).

Explanation of 'Cyber War' by Clarke and Knake

Clarke and Knake, in their book *Cyber War*, explain how cyber weapons work and how vulnerable America is to the new world of nearly untraceable cybercriminals and spies. They further reveal that successful foreign cyber espionage is already in the United States. They stated it has "penetrated the Pentagon, the control systems for U.S. electric power grids, and the defense industry." Although the United States has not yet received a full-scale cyberwar, hackers have already stolen information, including advanced research in aerospace, weapons systems, biotechnology, and engineering (Clarke & Knake 2010).

World Nations and Cyber Wars

First, cyber wars and attacks make up the world silent wars waged on nations by nations. *Second,* US security, including classified information, also has been under attack. The United States accused North Korea of a cyber- attack on their nation. Another was during presidential campaigns. For example, hackers got into the email system of the 2016 Democratic candidate, Hillary Clinton. Many believe the hack had a great deal to do with the outcome of the election.

Since the presidential election, there has been more talk regarding hacking. The US security system has accused Russia of a cyberattack on US democracy. However, among the US politicians, there seems to be a political split on the Russian cyberattack on the country. As of 2019, the investigation concerning this hacking is still in hot pursuit. Members of the Democratic Party feel that the involvement of the Russians caused them to lose the election. Regardless of who is right or wrong regarding this hacking into the securities of the United States, the fact is, cyber-attacks are a new method of warfare, and they will continue.

Third, as long as people live in a cyber world where computers are the most reliable mechanism for information-sharing and information- stealing, attacks will continue to influence countries' securities and businesses. Humankind has, and will look for opportunities to ignite strife, uprisings, rumors of wars, and actual war because there is a fundamental need to control or be in charge. Being in charge is one of the primary goals that nations have and will fight and strive to achieve.

Fourth, by waging such silent cyberwars with the use of computers, a nation can attack another by hacking into the computer systems. Countries that are engaged in cyber warfare can steal high-security information from vulnerable countries. Such theft gives those countries the upper hand. Nations will continue to use this form of warfare because they like to have access to another country's military systems and secrets. Enemy countries want to know how strong or weak another country can be and how to gain opportunities to threaten their existence. Many countries are seen by others as enemies and want them to be vulnerable and powerless.

Fifth, the year 2021 revealed that the United States' Government and many of its companies were breached. "The US intelligence says this massive hack was done by Russia and it went FAR BEYOND SolarWinds" stated the Daily Mail (2021). The Straits Times in an article "US intelligence says Russia likely behind hacking of government agencies' stated:

> The incoming administration of Democrat Joe Biden has already promised a response to the SolarWinds hacks. On Tuesday, the top Democrats on the Congressional intelligence committees underscored that need.
>
> "Congress will need to conduct a comprehensive review of the circumstances leading to this compromise, assess the deficiencies in our defenses, take stock of the sufficiency of our response in order to prevent this from happening again, and ensure that we respond appropriately," said Mr Adam Schiff, head of the House committee.
>
> Russian officials have denied involvement and did not immediately respond to questions on Tuesday (The Straits Times 2021).

CNN's Fareed Zakaria GPS and the Nations' Cyber Wars

On CNN's *Fareed Zakaria GPS*, revealed that a former top British spy stated, "The world is chaotic and dangerous today." The Internet has no global process for setting rules for usage. There is no real ability to deal with it in a controlled and perfect manner. *Zakaria GPS Documentary* on cyberwar also stated these other interesting points:

(1) In the cyber world actors are diverse (and sometimes anonymous),
(2) Physical distance is immaterial, and some forms of an offense are cheap,
(3) Because the Internet is for ease of use rather than security, attackers have the advantage over defenders,
(4) Technological evolution, including efforts to "re-engineer" some systems for greater security, might eventually change that, but, for now, it remains the case,
(5) The larger party has limited ability to disarm or destroy the enemy, occupy territory, or use counterforce strategies effectively,
(6) Cyberwar, though only incipient at this stage, is the most dramatic of the potential threats,
(7) Major states with elaborate technical and human resources could, in principle, create massive disruption and physical destruction through cyber-attacks on military and civilian targets.
(8) Responses to cyberwar include a form of interstate deterrence through denial and entanglement, offensive capabilities, and designs for rapid network and infrastructure recovery if deterrence fails.
(9) At some point, it may be possible to reinforce these steps with certain rudimentary norms and arms control, but the world is at an early stage in this process. (Zakaria 2016)

We can conclude that cyber warfare will continue as nations spy on other countries, for there is no control over the Internet. Therefore, as technology advances, each nation will continue its usage and try to know the strengths, weaknesses, opportunities, and threats that another country has. The nation that has greater resources, technologies and cyber professionals will advance and win the battles of cyber warfare. Such a developing country will continue to manipulate and control the other.

Also, cyber warfare remains an ongoing political problem for nations and their free elections. As seen within the United States, there seems to be no end to the "Russia Probe of 2016 Election." Today in 2019, there seems to be other interference by Russia regarding the 2020 US election, according to the news media.

The above information gives readers an idea of the unusual signs or events that are taking place in the sky. Now, we must investigate the signs that are taking place on the land.

Chapter 2

STRANGE AND UNUSUAL 'SIGNS' ARE OCCURRING ON THE LAND

The news media presented stories that show many signs which are taking place on the land today. For example, (1) *People of the World are Changing, Unusual Events of People Against Dictators and Questionable Leaders, Hatred among Humankind Is Increasing Worldwide,* (4) *Increased Incidents of Uprisings, Violence, Terrorism, and Wars,* (5) *Modern-Day Pirating,* (6) *Worldwide Terrorism,* (7) *Fleeing Refugees,* (8) *Unusual Plagues and Diseases,* (9) *Unusual Instability in Leadership and Government,* (10) *Unusual Signs of the Failure of World Economic Systems, and* (11) *Unusual Events among Animals of the Earth.* The chapter now investigates these eleven signs.

Land Sign No. 1: People of the World are Changing

A 2013 article in the *Economist* reminds world leaders that (1) people of the world are angry with politicians and crooked leaders. (2) People are forming solidarity with each other where there is corruption, inefficiency, and arrogance of those in power. (3) People are calling for freedom and they are rejecting world dictatorships. Social media has made it possible for people to revolt. It has also made the world a smaller place. (4) Today the youth are using the social media tool as a means for communication with the world, thus, forming common solidarity among the masses of people around the world. (5) Rebellion is everywhere, and change is in the air because people are changing.

Land Sign No. 2: Unusual Events of People Against Dictators and Questionable Leaders

Riotous Crowds

Regarding people's actions toward dictators and questionable leaders, this section of the chapter presents four areas for discussion.

First, many countries are rebelling against dictatorships, while some countries' people are sitting down and allowing such governments to ignite a form of dictatorship by their duly elected leaders. However, looking back at the last few years, people have broken the chains of tyrannical control and set themselves free from such rule.

Second, riots and protests have increased around the world. Protests have many different origins. For example, a 2013 article in the *Economist* stated the people protest for different reasons: "In Brazil, people revolted against bus fares; in Turkey against a building project; in Indonesia against high fuel prices; in Bulgaria, the government's cronyism; in the eurozone, against austerity; in Hong Kong for freedom; and in the Arab spring pretty much everything. Each angry demonstration is angry in its own way (the Economist, 2013).

In the United States, several groups protested against the results of the 2016 presidential election. They felt that Russia's hacking and interference in Hillary Clinton's email system rigged the election. Wikipedia stated:

1. protests have expressed opposition to Trump's campaign rhetoric, his electoral win, his inauguration, his alleged history of sexual misconduct and various presidential

actions, most notably his aggressive family separation policy.

2. With some protests, people have taken the form of walk-outs, business closures, petitions as well as rallies, demonstrations, and marches.
3. The largest organized protest against Trump was the day after his inauguration; millions protested on January 21, 2017, during the Women's March, making it the largest single-day protest in the history of the United States (Wikipedia).

Third, the United States had several uprisings since the onslaught of the Coronavirus Crisis. For example, heavily armed protesters gathered at Michigan's Capitol in Lansing to decry the 'Stay-At-Home Order' that was put into effect by its Democratic Governor Gretchen Whitmer. The protesters who gathered at the Capitol called themselves, 'Michigan United for Liberty. There were members within the groups who brought signs and compared the governor to Adolf Hitler. Not only did the protesters gather to protest but it was revealed that there were plans to capture the governor and had her executed. Many had nooses and Confederate flags. Some signs read, "Tyrants Get The Rope 'stated Abigail Censky of n p r news (2020).

Fourth, January 6, 2021, was another uprising and riot on the United States Capitol. The mob overcame the Capitol Police and entered the Capitol Building carrying flags, plaques, posters, and other paraphernalia. Many of the rioters also were armed. The Capitol doors and windows were broken open. Several people died because of this mob attack. The reason for the attack was over the fact that Mr. Biden won the election and the former US President Donald Trump refused to accept the outcome. He constantly claimed that the election was rigged and that he was the winner. Therefore, his supporters gathered in Washington DC by the thousands and overran the Capitol Building. The results of the riot are still under investigation by the FBI and other officials of the US Government.

Wikipedia (2021) presented an Overview of this story. Portions of the story are presented. Note:

The **storming of the United States Capitol** was a riot and violent attack against the 117th United States Congress at the United States Capitol on January 6, 2021. Part of wider protests, it was carried out by a mob of supporters of Donald Trump, the 45th president of the United States, in an attempt to overturn his defeat in the 2020 presidential election. The riot led to the evacuation and lockdown of the Capitol, and five deaths.

Called to action by Trump, thousands of his supporters gathered in Washington, D.C., on January 5 and 6 in support of his claim that the 2020 election had been "stolen" from him, and to demand that Vice President Mike Pence and Congress reject President-elect Joe Biden's victory...

Many became violent, assaulting Capitol Police officers and reporters, erecting a gallows on the Capitol grounds, and attempting to locate lawmakers to take hostage and harm. They chanted "Hang Mike Pence", blaming him for not rejecting the Electoral College votes, although he lacked the constitutional authority to do so. The rioters targeted House Speaker Nancy Pelosi (D–CA), vandalizing and looting her offices, as well as those of other members of Congress. Five people died from the events, while dozens more were injured.

The Federal Bureau of Investigation (FBI) has opened more than 170 investigations into the events, and indicated that many more are likely to come. Dozens of people present at the riot were later found to be listed in the FBI's Terrorist Screening Database, most as suspected white supremacists. Members of the Oath Keepers anti-government paramilitary group were indicted on conspiracy charges for allegedly staging a planned mission in the Capitol. Read the complete story on Wikipedia (see reference for information).

Fifth, Russia's Uprising took place on January 25, 2021, when Mr. Alexei Navalny on his return to his homeland of Russia was jailed by the government. Per Lucian Kim (2021), "Demonstrators in Russia braved extreme cold, police brutality and mass arrests, calling for the release of the opposition leader, who was detained last week shortly after returning to the country." Also, it was noted that these demonstrators were calling for the release and freedom for jailed Kremlin critic Alexei Navalny and freedom for their country after more than 20 years of

Vladimir Putin's rule (Kim, 2021).

Sixth, "protests rage around the world- but what comes next? Unrest is seemingly everywhere. News media stories give some indication for the reasons for the rebellions in Hong Kong, Lebanon, Chile, Catalonia and Iraq," stated the *Guardian.*

Note the reasons for the various protests in the mentioned areas:

1. In Lebanon, they are against a tax on WhatsApp and endemic corruption.
2. In Chile, a hike in the metro fare and rampant inequality.
3. In Hong Kong, an extradition bill and creeping authoritarianism.
4. In Algeria, a fifth term for an aging president and decades of military rule.
5. Russia, Serbia, Ukraine, and Albania have all seen major demonstrations.
6. UK, against Brexit.
7. France, with its yellow vest movement.
8. Spain, in the restive region of Catalonia.
9. The Middle East has convulsed with so much dissent that some are calling it a second wave of the Arab spring.
10. In South America, Brazil, Peru, Ecuador, Colombia, and Venezuela have experienced popular unrest. The list goes on.
11. The people of Iran rebel against "increases in fuel prices" (Aljazeera, 2019).
12. Protests in Russia because of the people's jailed leader Mr. Alexei Navalny (2021).
13. Riot on the United States Capitol because of the alleged rumor of an unfair Presidential Election between Donald Trump and Joseph Biden (2021).

Seventh, the data obtained from various studies "show that the amount of protests is increasing and is as high as the roaring 60s, and has been since about 2009," says Jacquelien van Stekelenburg, a professor who studies social change and conflict at Vrije University in Amsterdam. Note the findings:

1. Economic complaints do not drive all protests but widening gulfs between the haves and have-nots are radicalizing many young people in particular.
2. Oxfam stated in January that the world's 26 richest individuals owned as much wealth as the poorest half of the global population. Billionaires grew their combined fortunes by $2.5bn a day in 2018, while the relative wealth of the world's poorest 3.8 billion people declined by $500m a day.
3. Thousands of people have joined the rolling *Extinction Rebellion* protests. They feel that leaders are doing nothing about the ***climate crisis,*** and therefore, are content to leave the problem to the next generation.
4. Social media and the explosion of access to information is reordering hierarchies of knowledge and communication.
5. Authorities can fight back with extensive surveillance regimes or with digital blackouts of the kind India recently imposed in disputed Kashmir. But 20th-century power structures are under enormous pressure, analysts stated (Michal Safi and international correspondents, 2019).

The news media reminds people that many areas of the world experienced protests in 2019. The conflict was present then and even now in 2020. The conflict happens because of many different reasons. However, the real cause seems to be 'The Hate' within people's hearts.

Land Sign No. 3: Hatred among Humankind is Increasing Worldwide

History's Records of Hate among Humankind

In researching the history of wars and conflicts among humankind, the records show that this planet is barbaric and without human reasoning of any kind. The years from 1920 to 2012 seemed without peace among people. Note:

1. In 1920 there was the Kurdish-Iranian conflict.
2. In 1948, the Korean conflict took place, as did the Baluchistan conflict, the internal conflict in Burma, and the Israeli- Palestinian conflict.
3. In 1953 the Nigerian Sharia conflict and the insurgency in Northeast India took place.
4. In 1964 there was the Naxalite-Maoist insurgency in South Asia.
5. In 1969 the Islamic insurgency in the Philippines and the Papua conflict took place.
6. The insurgency in Laos was in 1975, and in 1978 the Kurdish- Turkish conflict started.
7. In 1982 the Casamance conflict of Africa occurred. Many people died needlessly in this conflict.
8. In 1987 the Lord's Resistance Army insurgency of Africa took place.
9. In 1989 there was an insurgency in Jammu and Kashmir in South Asia.
10. In 1994 the Cabinda conflict in Africa occurred.
11. In 1995, there was an insurgency in Ogaden, Africa.
12. The dissident Irish Republican campaign took place in 1998, and four years later, the insurgency in the Maghreb occurred in Africa.
13. The Shia insurgency in Yemen in the Middle East occurred in 2004, and in the same year, there was a conflict in the Niger Delta, Africa.
14. Four years later, in 2008, the South Thailand insurgency in Southeast Asia occurred. That same year the Cambodian-Thai border dispute took place.
15. In the following year, 2009, the insurgency in the North Caucasus in Europe and the South Yemen insurgency took place.
16. In 2010 there was the Yemen al-Qaeda crackdown in the

Middle East.

17. In 2011, there was the North Kosovo crisis.
18. In 2011 the Libyan factional fighting was also.
19. The Tuareg rebellion took place in 2012, as did the East DR Congo conflict and the Lebanese conflict in the Middle East.

From the various conflicts mentioned, note the number of people who lost their lives. For example:

1. In the 1964 Colombian armed conflict, 150,000–200,000 people died.
2. In the Afghan civil war in 1978, three million people died.
3. In 1991, 300,000–400,000 people died.
4. In the wars in Northwest Pakistan in 2004, 38,800 people died.
5. In 2006, during the Mexican drug war, 60,420 people died.
6. The Sudanese nomadic conflicts in 2009 took the lives of 5,641 people.
7. In 2011 during the Sudan internal conflict, 1,500 people died.
8. In the Syrian uprising of 2011, over 10,000 people died, and many more were destined to die as it continued.
9. In the 2011 insurgency in Iraq, 1,136–1,436 people died.
10. In 2012, the South Sudan border conflict took the lives of 1,538 individuals.

These conflicts and insurgencies account for millions of lives taken and as the years continued, many others have died. Why? Because of unrest, greed, hatred, selfishness, and the lack of love for one's fellow humans, people continue to kill each other.

Present Day Destruction of Humans by Each Other

First, besides these killings, other smaller-scale armed conflicts cause the death of human beings. In the modern conflicts in the Middle

East, the conflicts in North Africa, the active rebel groups, and the terrorist incidents, there will be many more deaths.

Second, Killing Continues in the Streets of Nations around the World. People killing people continues in the streets of nations with handguns, knives, and with the use of vehicles as another weapon. Some of the killings were because of drugs and other innocent victims. Note, the killing continues to occur.

Third, self-boasting, self-love, and covetousness seem to have taken over human touch and kindness. Take, for example, the following story of the killing on the metro train. In 2015, a man killed another on a subway train in Washington, DC, as others watched, and no one attempted to help to prevent the unnecessary death (The Washington Post 2015). Also, a forty-nine-year-old woman from Queens, New York; someone pushed her in front of a subway moving train in Times Square - she died (Santora 2016). Two employees of a New Jersey daycare center instigated Fight Club-style brawls between the toddlers and shared footage of the pint-sized pugilists on Snapchat (Degregory & Li 2015).

It seems as if humankind has lost the natural affection for one another as if love has become a lost word. Self-love appears to mean more to some people than loving another. Thus, increasing characteristics lend themselves to covetousness, boasting, pride, disobedience, unthankfulness, untrustworthiness, hate, failure to keep one's word, high-mindedness, and lovers of much pleasure. The strange thing about our times is that people who are engaged in unkindness or hurting someone, some people praise and uplift. But the people who try to live a life pleasing to the law and attempt to do good for and to others, we put down or degrade – they receive no praise. These are strange and unusual times.

Research Shows that Hate in the United States is Growing

First, Dastagir (2017) of USA Today, in an article titled The State of hate in America: In an America deeply divided, hate incidents appear to be increasing and growing more brutal, cited many incidents of hate. For example, the article spoke of racial hatred taking place with speech, symbols, and slurs. Hate groups are celebrating in rallies and with graffiti on walls and the personal property of the people to whom they direct the hate.

Second, the article noted this great divide that is in the United States. For example, the FBI tracked 5,818 hate crimes in 2015, which was an increase of 6.5 percent from the previous year. Hate groups in the United States rose for the second year in a row in 2016, and anti-Semitic incidents rose 86 percent in the first quarter of 2017. The FBI estimated there are 5,000 to 6,000 hate crimes a year in the United States. Other sources stated that Americans experienced an average of 250,000 hate-crime victimizations each year from 2004 to 2015. These statistics show that the incidents of hate could be higher since some responsible areas for recording hate crimes fail (Dastagir 2017).

Third, these are just a few stories about people in the United States who show no feelings for other humans. It seems the human touch has left people, and hate is all some people display as seen in 2019. What about the rest of the world where there is so much killing? The entire world is in a hateful mess, and there is a direct reason for it. Read on!

Land Sign No. 4: Increased Incidents of Uprisings, Violence, Terrorism, and Wars Continue

First, incidents of uprisings, violence, terrorism, and wars can all be discussed under the heading of violence. When it comes to the subject of violence, the United Nations relies on the Uppsala Conflict Data Program (UCDP) for such information. The UCDP is a university-based data collection program on organized violence. The program continuously gathers information and collects data, both quantitative and qualitative, on the world and its political systems about armed conflict and other types of organized violence. The collected data revealed that there is low-intensity conflict, military-involved conflict, political conflict, and conflict because of terrorism. Meaning, conflict, and violence are constant factors in the world.

Second, to sum it all up, the world and the nations are warlike. People today have no more natural love for each other because their mind-set is in a state of hate and war. Yet, among the people of the world, there are those who are trying to be effective. One such organization is Our Brothers' Keepers. This organization's main goals are to, (1) assist people with the deadly disease of HIV/AIDS; and (2), support youths to achieve real-life goals. The organization emerged from the story of Cain, who killed his brother, Abel. Therefore, the founders of Our Brothers' Keepers used this story from the Bible as the basis for their organization. Note Genesis 4:9, "And the LORD said unto Cain, 'Where is Abel thy brother?' And he said, 'I know not: Am I my brother's keeper?'" And he said, 'I know not: *Am I my brother's keeper?'"* The organization promotes that "We "when people talk about the pirates today, they consider them like the legends from the dark past.

Third, 'Modern Pirates' have changed a bit, but their activities and behavior are almost the same even nowadays." The news media identified in their stories piracy among the people of Somalia. Pirates have again taken to the seas, but the pirates today differ from the old sea robbers. They now have access to speedboats with radar and sonar. They have modern equipment for communication with other vessels. Modern pirates also have access to advanced weapons such as machine guns, torpedoes, and rocket-propelled grenades. The modern pirate also has access to satellite communications. Such advanced communications make it easy to follow and capture ships of other countries that are sailing in nearby waters. On capturing ships on the high seas, they hold the captain and crew hostage for ransom. The company's home office must pay large sums of money to have them released. This form of piracy has affected the economy of countries (The Way of the Pirates 2016).

Land Sign No. 5: Modern-Day Pirating

The Meaning of Pirating

The term pirating "can include acts committed in the air, on land (especially across national borders or about taking over and robbing a car or train, or in other major bodies of water or on a shore" (Wikipedia). However, pirating is another form of hate, that produces robbery or criminal violence, and terrorism.

The Pirating of Yesteryears

History records how many nations of Europe practiced the old piracy form of aggression in yesteryear. Pirates are those who engage in this kind of tyranny. History identifies pirates as a group of thieves and murderers. Because of pirates, people who lived in many West Indian Islands and the Americas lost their lives. Pirates even plundered the west coast of Africa, taking Black people into bondage and sold them into the slavery of then known as "New world colonization."

The coast of Africa and the West Indies became a focus for maritime piracy. History, depending on which country was writing it, recorded this form of theft as the colonization of the Americas and the West Indies. It was all classified as the discovery of new territory for whatever the invading pirate ships' flag flew. Although this form of piracy appeared to disappear, it continued in different forms among powerful nations. Even today, some nations that are more powerful than others will invade and take control over its people and their economics and freedom – that's piracy. However, today another form of piracy exists – *the modern-day pirates.*

The Modern-Day Pirating

In the last few years, some have gone back to the original form of piracy. A recent article was written on Modern Piracy (2016), agrees that "when people talk about the pirates today, they consider them like the legends from the dark past. Modern pirates have changed a bit, but their activities and behavior are almost the same even nowadays." The news media identified in their stories piracy among the people of Somalia.

Pirates have again taken to the seas, but the pirates today differ from the old sea robbers. Today pirates have access to speedboats with radar and sonar. They have modern equipment for communication with other vessels. Modern pirates also have access to advanced weapons such as machine guns, torpedoes, and rocket-propelled grenades. The modern pirate also has access to satellite communications. Such advanced communications make it easy to follow and capture ships of other countries that are sailing in nearby waters. On capturing ships on the high seas, they hold the captain and crew hostage for ransom. The company's home office must pay large sums of money to have them released. This form of piracy has affected the economy of countries (The Way of the Pirates 2016).

As of 2019, pirates continue their attacks in South America and in the seas off West Africa. The International Maritime Bureau's (IMB) states that the seas off West Africa are the world's worst area for pirate attacks (see the Bureau's map). According to the latest report (ICC, 2019), the pirates seized 75 vessels and crew as hostage onboard or kidnapped for ransom worldwide in 2019. Sixty-two of those captured were in the Gulf of Guinea – off the coasts of Benin, Cameroon, Guinea, Nigeria, and Togo. Worldwide, the International Chamber of Commerce's IMB Piracy Reporting Centre recorded 78 incidents of piracy and armed robbery against ships in the first half of 2019, compared with 107 incidents for the same period of 2018. Overall, 57 vessels pirates boarded successfully, representing 73% of all attacks (ICC, 2019). The other area of the world which is at risk for pirate attacks is in the seas of South America. (see the map below)

ICC International Maritime Bureau's Map >> Pirates' Hangouts

Land Sign No. 6:Worldwide Terrorism

There are other forms of terrorism that are not just on the high seas but may take place anytime and anywhere. Worldwide terrorism has become a big problem for the nations of the world. We could attach this form of terrorism to a person's religious and political teachings. The

Becker–Posner Blog on Terrorism and Economic Development stated that:

> Terrorism is a political phenomenon, and the demand is driven mainly by political grievances, real or imagined. Often the grievances are related to foreign occupation. France in Algeria; the British in Palestine; now the Israelis in the West Bank; the United States in Iraq (and earlier in the Philippines) --though in the case of ... terrorism, the major factor seems to be the Western "presence" in the Middle East, rather than a foreign occupation; even Israel's occupation of the West Bank seems a subsidiary factor. The Baader-Meinhof gang in West Germany, the Red Brigades in Italy, and Aum Shinrikyo in Japan are examples of terrorist groups' unrelated to foreign occupation. But it is the existence of grievance that is key, and often- -probably typically--the grievance is political rather than economic. (Becker & Posner 2008)

Besides the above, some groups, whether religious or political, their followers are so indoctrinated that they will give their lives for the cause. Therefore, in the name of a god or a political ideology, will carry out an event that will cause the loss of their lives and those are the target of the terrorist event.

Terrorism's Effect on Countries around the World Terrorism is a major factor in countries around the world. It has plagued the world and added to the downfall of the world's economic system. The blog by Becker & Posner infers that people who live in their countries of origin, which another country occupies, have feelings of always being controlled. Therefore, they form religious or political groups and react with acts of terrorism. Such acts of terrorism add a financial burden to the country that these terrorists attack. One example of this is the 9/11 terrorist incident in New York City that leveled the towers and killed many people. As a result, the United States spent millions of dollars on new procedures at entry points, such as airports (Abadie 2004).

After the attack on the United States, all around the world—in England, and many of the allies of the United Nations—there was an upgrading of security that continues to this day. Countries spent millions of dollars to improve and prevent terrorist attacks which has a tremendous effect on the economic systems of such countries.

Terrorism is Still Alive and Continuing in Countries around the World

Collectively countries have killed three or four of the main leaders of terrorism. However, terrorist incidents have not ceased. For example, in 2019, from January to November, Wikipedia lists terrorist incidents for each month. Note some of the cases:

- In January 2019, there were 7 incidents by shootings, car bombings, and suicide bombers. Approximately 266 people died, and it injured about 293 people.
- In February 2019, there were 11 incidents. Approximately 313 died and it injured 180. In one case the injured number is unknown.
- In March 2019, there were 10 incidents of terrorists' attacks. Approximately 276 people died, and it injured about 142 people.
- In April 2019, there were 11 attacks by terrorists. Approximately 376 people died, and it injured about 5856 people.
- In May 2019, there were 11 terrorist attacks that killed approximately 202 people and wounded about 499.
- In June 2019, terrorists did 16 attacks. Approximately 540 people died, and it wounded 820 people.
- In July 2019, there were 9 terrorist attacks. Approximately 202 people were killed and 297 injured.
- In August 2019, there were 8 terrorist attacks. Approximately 152 people died, and it injured 258 people.
- In September 2019, 7 terrorist attacks took place. Approximately 147 people died, and it injured 254 people.
- In October 2019, 9 terrorist attacks took place. Approximately 152 people died, and it wounded 83 people. In this attack, the wounded are unknown.
- In November 2019, the terrorist attacks were 7. Approximately 168 people died, and it injured 179 people.

- In December 2019, there were 11 terrorist incidents. Approximately 259 people were killed, 262 people were injured, and for many others, the count was unable to get.
- In January 2020, there were 7 terrorist attacks worldwide. Approximately 217 people were killed and 262 were noted to be injured. Many others were injured but there is no count.
- In February, as of 2-6-2020, there were 3 incidents. Nobody was killed but 12 people were injured.

The methods for the terrorists' attacks were acts of shooting, car bombs, suicide bombing, hand grenades, etc. These attacks took place in various parts of the world. Mostly in Syria, Afghanistan, Thailand, Burkina Faso, Mali, India, Indonesia, Germany, France, Iraq, Turkey, West Bank, United Kingdom, Norway, Egypt, Texas–US, Gilroy–US, Pensacola–US, Jersey City–US, Louisiana–US, California–US, Monsey–US, Nigeria, Somalia, Philippines, Ethiopia, Dallas, Pakistan, Saudi Arabia, Australia, Democratic Republic of the Congo, Sudan, Lebanon, Israel, Nepal, Tajikistan, Pakistan, Cameroon, Poway, Sri Lanka, Mexico, Netherlands, New Zealand, Spain, Iran, the Maldives, Niger, the Democratic Republic of the Congo, Faso, Kenya, and Yemen. Incidents in some countries were over one time (Wikipedia).

Terrorism is still alive and killing and wounding innocent people. Such hate will continue according to Biblical Prophecies. The hate and killing are causing many people to flee their homelands, looking for a place of refuge – increasing the refugee problem.

Land Sign No. 7: Fleeing Refugees

According to the UNHCR – The UN Refugee Agency, "Every minute in 2018, 25 people were forced to flee." Filippo Grandi, the United Nations High Commissioner for Refugees states, "What we are seeing in these figures is further confirmation of a longer-term rising trend in the number of people needing safety from war, conflict, and persecution." Among these groups of refugees, internally displaced people, asylum seekers, unaccompanied and separated children, stateless people, and others. The number of these displaced people has increased by 2.3 million in 2018. By the end of such a year, almost 70.8 million individuals were forcibly displaced worldwide because of persecution, conflict, violence, or human rights violations. Note the graft below which shows the refugee increases. The influx of refugees worldwide is increasing. The fact that war, conflict, and persecution are not stopping, means the refugee population will continue.

As these acts of violence by wars and other uprisings continue, they are causing an insurmountable number of refugees all over the world. The refugees have imposed themselves on other countries. Many countries that cannot even help themselves now face the problem of caring and providing for refugees. To help in the refugee situation, world systems for humankind have set up camps along various countries' borders where refugees may have a temporary secure place to lodge. However, in many of the refugee camps, the people do not have food and water. Also, sanitation problems increase, and diseases are on the rise within these camps.

Besides the shortages of food, water, medical supplies, increased diseases, finding refugees a permanent place to live is another problem the world is facing. Countries, where the refugees are fleeing, must accept or reject them. Some nations have placed border security and, sometimes, have deployed their military to prevent the inflow of refugees to their country. So, today around the world, many leaders are facing political backlash regarding refugees, which is becoming an international problem.

The News Media Share International Refugee Political Problems Nations Faced

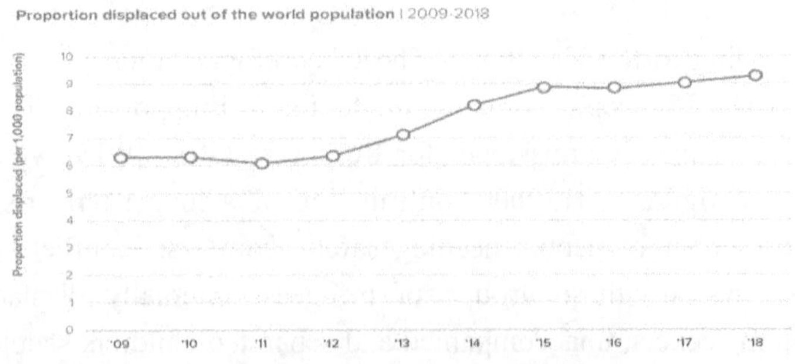

Graft: Increasing Numbers of Refugees

First, during the 2015 refugee problem, the United Nations Refugee Agency wanted the United States to take in more of the refugees fleeing Syria's civil war. The then US President Obama, under pressure from the international community, decided that the United States would take in around ten thousand refugees from war-torn Syria, but the number did not satisfy the UN Refugee Agency, they preferred a greater number. However, at that same time, Germany took some 800,000 refugees. (Brannon 2015)(see graft)

Second, the Los Angeles Times described the refugee problem within the politics of Israel. The prime minister Benjamin Netanyahu rejected a call to host refugees from Syria and elsewhere. There was much political conflict regarding the prime minister's decision by the opposition. There was a feeling that Israel should be more sympathetic to the plight of the refugees. People felt that Jewish history demands compassion and not to live in the "world's silence and remain indifferent" to the carnage in Syria and the refugees' plight (Sobelman, 2015).

Third, a *New York Times* article stated that the Arab nations of the Persian Gulf have some of the world's highest per capita incomes. Their leaders speak passionately about the plight of the Syrians, and their state funded news media cover the Syrian civil war without ceasing. As millions of Syrian refugees languished elsewhere in the Middle East, and many risked their lives to reach Europe or died along the way, Gulf nations agreed to resettle a surprisingly small number of refugees (Hubbard 2015).

Fourth, the San Jose Mercury News (2015) story was about Germany's leading role within the European Union and the twenty-eight-nation bloc. The story stated that the prime minister failed to persuade his Czech, Slovak, Polish, and Hungarian counterparts to drop their objections to a proposed EU-wide quota system. The problem to help immigrants already overburdened EU nations created much political confusion over the refusal to receive refugees among the European Union bloc of nations and its leadership capabilities.

Unsanitary Living Conditions of Refugees and the Influx of Diseases and Death

Unsanitary living conditions also increase the influx of diseases and death. The ReliefWeb released an article, "Refugee Camps Spread Life Threatening Diseases." The article stated that severe food and water problems had turned the refugee camps into breeding grounds for a range of life-threatening diseases. Health officials cited outbreaks of cholera, malaria, and acute jaundice, coupled with widespread malnutrition, threaten the lives of many who thought they would be safe when they fled from their homes to refugee camps. In some camps, there were also outbreaks of hepatitis E, a viral infection transmitted by contaminated food and water. This disease hit hard for people between the ages of fifteen and forty (Sinha 2010).

Climate Change Will Become a Major Cause for Future Refugee Problems

Apart from the present problems described which refugees encounter, the other problem is that of Global Warming/Climate Change. Climate Change is and will continue to affect low lying areas of the world which will influence the migration of refugees.

A report by John Podesta (2019) on "The climate crisis, migration, and refugees," showed that cyclones and subsequent flooding damage destroyed acres of crops and demolished infrastructure causing people to scramble for temporary sites to live. These storms have created a multitude of critical issues that the international community must confront. These are: (1) large-scale human migration because of resource scarcity; (2) increased frequency of extreme weather events; (3) intensifying intrastate and interstate competition for food, water, and other resources, particularly in the Middle East and North Africa; (4) increased frequency and severity of disease outbreaks; and (5), increased U.S. border stress because of the severe effects of climate change in parts of Central America. These challenges are serious, but the scope and scale of human migration because of climate change will test the limits of national and global governance and international cooperation (Podesta, 2019).

Climate Change will cause people to move to higher ground for safety. The move will create another influx of refugees and the challenge for the world and refugees will continue. People may gather additional information on Climate Change and refugees from these three scholarly articles, (1) *The first climate refugees? Contesting global narratives of Climate change in Tuvalu* by Carol Farbotko, (2012), (2) *Turning the Tide by Ted Williams*, and (3), *Confronting a Rising Tide: A Proposal for a Convention on Climate Change Refugee* by Bonnie Docherty and Tyler Giannini. Land Sign No. 8: Unusual Plagues and Diseases

Land Sign No. 8: Unusual Plagues and Diseases

The Facts and Stats of Human Diseases

HIV/AIDS, the Ebola virus, heart disease, hypertension, cancer, cholera, malaria, and acute jaundice, malnutrition, hepatitis E, smallpox, syphilis, typhus, tuberculosis, and many other diseases are wreaking havoc in some countries. The most recent attack of the Ebola virus brought panic to the world. Recently, the increase of refugees has led to the potential for diseases transmitted in the camps. Also, the Centers for Disease Control and Prevention (CDC) warns of the Zika virus, transmitted by mosquitoes in South America, as a new threat and cause for concern (LaMotte 2016).

Many more mosquitoes, however, are carrying other dangerous diseases. Many of the older diseases, such as smallpox and the bubonic plague, have killed millions of people throughout the world. HIV and AIDS are among the newest and deadliest diseases. Per the World Health Organization, it is unknown where HIV originated, but it appears to have moved from animals to humans. Science states that HIV arose from a less harmful virus that mutated and became more virulent. As of 2013, about 1.3 million persons in the United States were living with HIV or AIDS and almost 110,000 in the UK (The National Aids Trust). Approximately thirty-five million people worldwide are living with HIV (World Health Organization).

Despite the efforts in many countries, awareness and prevention programs have not been effective enough to reduce the number of new HIV cases in many parts of the world. Countries that have high mobility of men, poverty, and sexual mores are more at-risk. In countries such as the UK, there is no significant decline in certain at-risk communities. Health records for the year 2014 showed the greatest number of new diagnoses in gay men, the equivalent of nine being diagnosed a day (Public Health England 2015). The World Health Organization estimated that in 2013, thirty-five million people were living with HIV/AIDS and 1.5 million people died from the disease (Global Health Observatory Data 2015).

Another statistic states that there are thirty-seven million people infected with HIV. Two to six million of these are under the age of fifteen. In 2014, the HIV virus infected about two million people. Among these newly infected people, 220,000 were under the age of fifteen. Science estimates that every day, 5,600 people contract HIV, over 230 every hour. In 2014 1-2 people died from AIDS. Since the beginning of the pandemic, nearly seventy-eight million people have contracted HIV and close to thirty-nine million have died of AIDS-related causes (amfAR Making AIDS History 2015).

The Ebola outbreak, which was the twenty-sixth outbreak since 1976, started in Guinea in March 2014. The World Health Organization (WHO) warned that the number of Ebola patients could rise to twenty thousand and said that it spent $489 million to contain Ebola within six to nine months (Wikipedia). Per Health Intelligence, since the onset of the disease in 1976, 280 people have died in the Democratic Republic of Congo. In that same year in Sudan, 151 people died. In 1977, people perished in the Democratic Republic of Congo. In 1979 in Sudan, there were twenty-two more deaths from Ebola. In 1994, thirty-one died in Gabon. In 1995 there were 254 deaths in the Democratic Republic of Congo.

As there is no proven treatment for Ebola, the total costs for medical items were proportionally lower to purchase compared to other diseases that need more expensive drugs and equipment. To bring Ebola under control requires more than caring for patients. For example, outreach activities such as contact tracing, health promotion, and disinfection of contaminated houses also represented a fundamental part of MSF's activities, with teams working to detect and prevent the virus within the communities (Medecins Sans Frontieres (MSF) - Doctors Without Borders). So far Ebola has killed over 11,000 people. Statistics are given by World Health Organization, and as of 27 March, Ebola cases and deaths are Guinea, 3811 cases and 2543 deaths; Liberia, 10,675 cases and 4809 deaths; Sierra Leone, 14, 14,124 cases and 3956 deaths; Italy, 1 case and 0 death; Mali, 8 cases and 6 deaths; Nigeria, 20 cases and 8 deaths; Senegal, 1 case and 0 death; Spain, 1 case and 0 death; United Kingdom, 1 case and 0 death; United States of America, 4 cases and 1 death. The total cases are 28, 646 and total deaths are 11,323 (World Health Organization, 2016).

From 1996 to 2012, the Ebola virus infected many African people. For example, In 1996, sixty-six people died in Gabon, and in the same year, individuals died in South Africa. In the year 2000 in Uganda, 224 people died from Ebola. Fifty-three people died from HIV in 2001 in Gabon and forty-four in the Congo.

In 2003, the Ebola virus had infected 128 people in the Congo. In that same year in the Congo, 29 people died of the HIV virus. In 2004, seven people died in Sudan, and in 2005, the Congo had ten deaths. In 2007 there were 187 deaths in the Democratic Republic of Congo. In that same year, there were thirty-seven deaths in Uganda. In 2008 the Democratic Republic of Congo had fourteen deaths.

In 2011 Uganda had five deaths caused by the Ebola virus. In 2012, there were twenty-nine deaths in the Democratic Republic of Congo and twenty-nine deaths in Uganda. The year 2014 appeared to put new emphasis on Ebola. For example, starting in the country of Guinea, 1,607 people died. Another 2,582 persons died in Sierra Leone, forty-nine died in the Democratic Republic of Congo, six in Mali, eight in Nigeria, and in Liberia, 3,384 people died from Ebola (Martiner 2014). For 2014, 7,636 people died in Africa from Ebola.

Facts: Diseases that Still Plague the World as of 2019 and Continues into the Next Centuries

According to the World Health Organization (WHO), infectious diseases are still threatening the inhabitants of the world in 2019. Disease problems confronting the world are:

1. Global Influenza Pandemic
2. Antimicrobial Resistance
3. Ebola and high-threat pathogens, including disease X
4. Vaccine hesitancy
5. Dengue
6. HIV and many others.

During 2019, there were disease outbreaks during the month of October. For example, (1) measles in Lebanon, (2) Middle East respiratory syndrome coronavirus (MERS-CoV) in the Kingdom of Saudi Arabia, (3) two outbreaks of Ebola virus disease in the Democratic Republic of Congo, (4) Cholera in the Republic of the Sudan. For the month of November, (1) Dengue fever in the Republic of Sudan, (2) Yellow fever in Bolivarian Republic of Venezuela, (3) Ebola virus disease in the Democratic Republic of the Congo, (4) Dengue fever in Pakistan, and (5) Rift Valley Fever in the Republic of the Sudan.

However, there are other existing diseases such as Chikungunya, Cholera, Crimean-Congo hemorrhagic fever, Hendra virus infection, Influenza infections, Lassa fever, Marburg virus disease, Meningitis, Monkeypox, Nipah virus infection, Plague, SARS, Smallpox, Tularaemia, Zika virus disease and others (World Health Organization (WHO) 2019).

7. CORONAVIRUS – COVID-19 PANDEMIC

First, as of January 2020, a former virus mentioned in the October 2019 group, the Coronavirus appears to have resurfaced. It is making havoc in China and appears to be spreading worldwide. This plague of "Coronavirus" which spreads from human to human, by a handshake or sprays from a cough. So far, a top Chinese government-appointed expert says a mysterious respiratory illness has killed at least four people. He also stated that humans transmit this virus. This became a great concern for nations and people. New statistics show that the virus has already infected many people in China. Also, some people who may carry the virus might not show symptoms (New York Times, 2020). However, many nations are taking precautions at airports and other entry points.

Second, January 30, 2020, Russia closed its borders near China as a precaution for this virus. As of February 12, 2020, the Coronavirus has killed more than 1,100 people, mostly in China, and a total of 44,653 confirmed cases. Also, it has spread to London, UK, was discovered on a cruise ship in Japan, and there have been alarms about the virus in the United States. There seems to be no stopping the virus.

Third, also on January 30, 2020, the Coronavirus or as was later branded became a Pandemic with the name 'COVID-19'. The pandemic has done much damage to countries financially, politically, and has caused the death of millions of people in the world. In addition, many countries had to close down much of their activities. Many businesses such as restaurants and other recreational accommodations had to close down which resulted in the closure of many of the businesses. Also, schools throughout the world were also closed and many countries are still trying to figure out how to have schools opened again.

Fourth, per the World Health Organization (WHO) they classified this as an infectious disease. Note:

> Coronavirus disease (COVID-19) is an infectious disease caused by a newly discovered coronavirus.
> Most people infected with the COVID-19 virus will experience mild to moderate respiratory illness and recover without requiring special treatment. Older people, and those with underlying medical problems like cardiovascular disease,

diabetes, chronic respiratory disease, and cancer are more likely to develop serious illness.

The best way to prevent and slow down transmission is to be well informed about the COVID-19 virus, the disease it causes and how it spreads. Protect yourself and others from infection by washing your hands or using an alcohol-based rub frequently and not touching your face.

The COVID-19 virus spreads primarily through droplets of saliva or discharge from the nose when an infected person coughs or sneezes, so it's important that you also practice respiratory etiquette (for example, by coughing into a flexed elbow).

Stay informed:

- Protect yourself: advice for the public
- Myth busters
- Questions and answers
- Situation reports
- All information on the COVID-19 outbreak

(see: https://www.who.int/health-topics/coronavirus#tab=tab_1

The information given by WHO was to assist people in learning about this new virus that was and is devastating the world. However, this organization received much criticism regarding giving out information that could assist with the fight against the pandemic. Nevertheless, the pandemic continues to attack the countries of the world and has continued its attack.

Fifth, in the United States, it seems that the country was hit hardest with the virus. For example, there were times when there was not enough protective gear for the health workers and there were not enough ventilators and ICU beds for those who were seriously sick from the virus. In addition, the death rate kept going higher and higher as time went by. So far, according to the statistics of the United States, the Center for Disease Control and Prevention (CDC) states:

> Rates of #COVID19 cases are widespread across the United States. As of January 24, over 25 million total cases of COVID-19 have been reported to CDC. Help slow the spread in your community: Wear a mask. Avoid crowds. Stat 6 feet apart. (CDC).

As of the above statistics given by CDC, as of January 29, 2021, in the United States, the Coronavirus Cases are; 26,453,428 and the Deaths are: 446,230. The stats are rising at a rate measured in seconds.

Sixth, the good news is, there are two vaccines available within the United States and the World. Also, there is another that will be out soon. However, within the United States there seems to be a problem with people getting the needed vaccines due to short supply. However, since there was the change of the US Administration, it appears that this problem will soon be mended. But people are still dying from the disease, and many are unable to get the vaccine. In fact, the majority of those who are heavily affected are Black people and people of color. Not that the disease can see color or ethnicity, but these are those who have been 'dished out' small amounts of medical care and assistance. These people are classified as the 'disenfranchised' "marginalized', excluded, or alienated.

Seventh, science is discovering new methods to fight the mutations because new statistics are still giving threatening data. For example, it is predicted by the CDC that 479,000 to 514,000 total COVID-19 deaths in the United States by February 20, 2021. However, it is also predicted that over the next four weeks, the number of newly reported deaths per week will likely decrease in 10 jurisdictions, which are indicated in the forecast plots. But future reported deaths are uncertain or predicted to remain stable in the other states and territories (CDC).

Eighth, it would appear that soon there will be help through the vaccines to give a feeling of hope. However, the virus is mutating and producing additional strains. Some of the variants appear to be not controlled by the newly made vaccines, but the vaccine push is being made by the scientists. Note: Dec. 13, 2020 -- Two vaccines have been approved for use in the United States (Pfizer and Moderna) a third and fourth are coming soon (AstraZeneca and Johnson & Johnson). Hopefully, these will help to save people's lives.

The virus is spreading faster than the author of this book can record it. Listen carefully, the above is a ***'SIGN'*** that diseases are on the rise and there are hardly any cures for them. The world seems to be undergoing an increase in *'pestilence.'* Jesus Christ while on earth, predicted that such occurrences of pestilences or disease, famines, earthquakes, wars, hate ,and deception will be 'Signs' of the *'End Time'* before His Return to the earth. Are we experiencing *'Signs of the End Time'?* This will be discussed in upcoming chapters of this book. So, let us continue with the strange and unusual signs that are taking place on the earth.

Land Sign No. 9: Unusual Instability in Leadership and Government

Strange and unusual events are taking place among governments throughout the world. Governments seem to fail, and many former leaders have lost their hold on political power. Some of these leaders lost both their position and life.

"People are Looking for a Leader Who Will Run With the World"

The Effects of Poor Leadership

First, poor leadership has produced the following problems in many governments around the world: (1) instability and the quick turnover of leadership, (2) a weak economic growth, (3) failure of financial systems, (4) lack of employment, and (5), sometimes, uprisings among people. Together these problems have a significant effect on the country and the people. An article titled "Political instability and economic growth" by Alesina, and colleagues (1992) reminds people that, (1) 'Political instability' causes governments to collapse, and (2), affects economic growth and can lead to poor financial performance, government collapse and political unrest. Also, problems generated by poor leadership leave the people under such leadership in desperate need and search for a global leader.

Second, today among the nations of the world, people are in this quest for a global leader. They are looking for a leader who will (1) take care of all the world's ills of economic difficulties, (2) human rights, and (3), the environment with its climate change problems. Concerning such global leadership, an article by Peter Blair Henry (2013) states that a world leader will be able to take care of Economic difficulties, environmental concerns, poverty, and human rights. A global leader would understand issues that cut across national boundaries, affecting individuals and entire populations around the world. He or she will understand how to make and implement decisions that have the global good in mind (Henry 2013).

The World is Gearing Up for Global Leadership

The many world problems appear so great that people see leadership among each country unable to solve the world's ills. Therefore, the world is already gearing up for global leadership. For example, the World Economic Forum operates a Global Leadership Fellows Program (GLEP). The program takes a step towards finding the next generation of leaders. Also, the program addresses the future of the management and governance of the world.

The GLEP aims to train and develop leaders. The principle underpinning is to influence, inspire, and work within and across cultures and organizations, while holding its capacity to motivate and develop people around them. Its core strategic aspirations are: (1) to foster visionary leadership; (2) to promote greater accountability towards the forum's communities and stakeholders; (3) to be constantly ahead of conventional thinking; (3) to instill impeccable intellectual and moral integrity; and (4), to inculcate the forum's commitment to serving the public interest.

The GLEP Organization has three pillars which are the foundational structure for world leadership.

First, at the most basic level, consistent with the mission of the World Economic Forum, which is "committed to improving the state of the world."

Second, the system's mindset is for global governance or world governance, where political actors will try to solve the world's ills, and people will join and follow the leadership that the world wants.

Third, to accomplish the goals, there is a world-leadership school curriculum which trains children from kindergarten to 12 grade. The idea is to present a mastery of skills where students will use skills to have an impact on the world. Students can make meaningful connections in the world and act – the key steps to clarifying purpose (World Leadership School, 2019).

Based on all the above activities of learning, the program hopes the world will one day have a Global Leader. However, as the program is trying to discover a global leader, the world economy continues on a decline.

Land Sign No. 10: Unusual Signs of the Failure of World Economic Systems

The World Economy of Yesteryears

Economies are collapsing worldwide. A brief analysis from the Council of Foreign Relations showed the global markets are dropping, banks are imploding, oil prices will increase, and doing business is sucking cash out of consumers' wallets. Countries like China are sitting on greenback reserves. Despite interest cuts, firms and hedge funds cannot borrow to pay off debts, companies are selling off other assets, and the spreads of losses are global, across sectors. This process will drag down markets worldwide (Teslik 2007).

Per Caballero and Simsek (2009), the Financial Times reveals a risk in giving credit today. As asset prices implode, more banks become distressed because of the likelihood of being hit by an indirect shock. This complexity brings about confusion and uncertainty. Relatively healthy banks and potential asset buyers become reluctant to buy. They fear becoming embroiled in a cascade they do not control or understand. Also, the Feds play a critical role in increasing and reducing interest rates.

A 2009 article in the Wall Street Journal on "The Great Recession" revealed that the Great Recession began as a national recession in the United States in December 2007. It only met the International Monetary Fund (IMF) criteria for being in a global downturn, but by the year 2009, Europe was in crisis (Wikipedia).

By late 2009, there was an increase in problems relating to government debt, banking, economic growth, and competitiveness. These crises became so bad that some countries had to bail out other countries who were having financial difficulty. Countries such as Greece, Ireland, Portugal, Spain, and Cyprus are some of these countries that needed intervention for survival (Shambaugh 2012).

The United States was not exempt from the crises that Europe faced. There were many bailouts in the United States because of the swirl of negative financial crises and declining economics. The largest bailout was the US insurers, the American International Group (AIG). We know this company as the US financial flag carrier. It had seen its shares pummeled following a series of credit downgrades. To have AIG go down would have been very negative to the United States (Council on Foreign Relations 2002).

Within the United States, the economy also took a dive, and the country had a mild recession. As of 2014, the country was still trying to dig itself out of such an economic disaster. Because of the recession in the United States, some of the major industries needed a bailout. For example, in November 2008, the US government had to give loans to the auto industry (GM, Chrysler, and Ford) to save these companies from their financial problems. Congress rejected the first request but after the huge layoffs took place, it forced Congress to allow the bill to pass.

Besides the bailout of the auto industry, the banking industries also were in crisis. On April 18, 2011, for the first time since the beginning of this rating system in 1860, the S&P, a US-based rating agency, issued a "negative" outlook. On July 16, 2011, the Egan-Jones rating company cut the US rating from 'triple-A' to 'double-A' rating. On August 5, representatives from S&P announced the company's decision to give its first-ever downgrade to US sovereign debt, lowering the rating one notch to AA+, with a negative outlook, per a press release. Both Democratic and Republican politicians criticized S&P's decision and placed blame with the other party. Few blamed themselves, despite a bi-partisan congressional responsibility for passing budget deficits from 2002 onward (Congressional Budget Office 2011).

Many Americans believed both Neil Barofsky's Bailout and Sheila Bair's Bull by the Horns—that the US government cared a lot more about saving the incumbent banks and bankers than it did about helping regular Americans. Thus, many Americans believe that the government had politically rigged the country's rules for a few connected financial speculators (*Economist,2012*).

The '*growth map'* of the global economy is relatively clear. The US is in a partial recovery … and lagging employment. Europe is barely above zero growth, with significant variations among countries. China is leveling off and other developing countries preparing for higher interest rates" (Spence & Leipziger 2013).

The World Economy Today

Today, the United States has been coming back from the recession. An article presented by No Labels (2017) described how the United States bottomed out in 2008, leaving the country in its worst financial state since the Great Depression. However, since 2017, there has been economic growth of its GDP 3.3 percent and the unemployment rate has decreased. However, economic growth appears to be beneficial to the rich. Not all Americans are 'basking' in economic growth as it seems. This is America but what about the economic growth of the rest of the world?

The question today is, 'In which direction the world economic system is heading?' According to the GDP growth of the world, between 1961 to 2018 there has been a steady decline. (see the attached World Bank Organization's data indicator:

https://data.worldbank.org/indicator/NY.GDP.MKTP.KD.ZG).

A report by the World Economic Forum first states that, "The world economy is heading for a crossroad." The organization also points out that there are Global economic imbalances, and in looking at the "Future of economic progress," Millennials earn less than past generations." Second, the European Investment Bank will stop funding fossil fuel projects by the end of 2021. Third, the manufacturing of motor vehicles is bracing itself for another major shake-up with the rise of electric vehicles that will influence world economics.

Fourth, the disparity of income between regions in countries is persistent and has risen over the past 15 years. Therefore, social and economic inequality is becoming a problem for world economics.

Land Sign No. 11: Unusual Events among Animals of the Earth

Photo: Some of the World's Animals who are Experiencing Unusual Events

Massive numbers of animals are dying off. The news media also share incredible stories of activities among the animals around the world. Animals such as sheep, pigs, cows, goats, deer, dogs, cattle, horses, donkeys, bats, bees, worms, and many others are dying off in massive numbers. Some countries have losses in the thousands, and there seems to be no letting up of these deaths. The news media have labeled the deaths as strange, unusual, mysterious, and as baffling to scientists, as there is no known reason. (See appendices at the back of the book and see the amounts of animal mass deaths up to 2019).

Animal deaths remain a mystery most times. However, science claims that many of the cases have been because of global warming's climate change. For example, the fires of Australia have killed over a billion animals, and their deaths are still occurring. Likewise, throughout the world, other storms are associated with Global Warming/Climate Change.

Chapter 3

Strange and Unusual 'Signs' are Occurring in the Waters

Just as strange and unusual signs are occurring in the sky and on the lands of the earth, there are also unusual signs in the waters of the world. In 2012, there were 465 water and sea animal deaths in sixty-seven countries. Among these deaths were fish and other creatures that live in the water. (For a full listing of the deaths of fish and other water animals from around the world, see the appendices at the back of the book.) The news media present the deaths of these animals as mysterious, weird, unknown, and catastrophic.

Freshwater and Sea Animal Deaths

There are so many unusual deaths of water animals occurring in the waters around the world, the author presents a few of the unusual and mysterious deaths that took place from 2012 through 2015.

The Water Animals' Deaths in 2012

On March 13, 2012, tens of thousands of salmon died in Marlborough, New Zealand (Marlborough Express 2012).

On June 6, 2012, in the United Kingdom, people found 5,000 fish dead in a lake (The Sentinel 2012).

On August 9, 2012, there was a mass death of shrimp in Vietnam (VICE News 2016).

On August 30, 2012, and again on September 16, citizens of Idaho found thousands of whitefish dead in the eastern Idaho river. Scientists concluded that kidney disease killed thousands of fish in 2012 (Earth Changing Extremities 2016).

On December 17, 2012, people in Iceland found thousands of herring dead in the water. They called this case mysterious. The Icelandic Marine Research Institute ruled that infection killed the herring (Iceland Review 2012).

The Water Animals' Deaths in 2013

Mass deaths of water creatures continued into 2013, reaching 798 incidents of mass deaths in ninety-three countries (End Times Prophecy 2013). These deaths were like those described in 2012.

An article written by Snyder in August 2013 captures the full story of strange and unusual deaths of water creatures. The writer stated that millions upon millions of fish suddenly died in mass-death events all over the world, and nobody seems to know why it happened. He also listed the mass deaths incidents for the year 2013. In the months of July and August, for example, thousands of fish deaths took place in China, Mexico, the United States, Canada, Denmark, Indonesia, France, Italy, England, Thailand, Korea, Belgium, Russia, Sweden, Romania, Alaska, Taiwan, the Czech Republic, Finland, and Pakistan. A large number of fish deaths occurred in different areas of each country, both in freshwater and the sea, and they occurred during a period of less than one month.

Why did this happen? Could they have done anything to stop it? The author of this article asked for comments from readers who wanted to share their opinions about the cause, but no readers responded.

The Water Animals' Deaths in 2014-2016

From 2014 through the beginning of 2016, the news media gave records of water animal deaths that were given with similar descriptions as the previous years. (For additional information, see the appendices.) These incidents of water creature deaths continue almost like a daily or weekly routine.

On April 13, 2015, a Chinese village awoke to one hundred tons of dead fish floating in their local pond (Perry 2015).

On April 21, 2016, in Vietnam, tons of fish, including rare species that live offshore and in deep water, the people of Vietnam discovered on beaches along with the country's central coastal provinces. Officials stated they'd seen nothing like it (France-Presse 2016).

The Water Animals' Deaths in 2019

As of 2019, the above photo shows the thousands of fish that mysteriously died and are floating on the water (see the appendices for additional information). In many of these cases, there seemed to be no clear answer, so the media labeled them as mysterious, unusual, unbelievable, or phenomenal. However, these strange and unusual deaths stunned the locals, baffled scientists, and caused a shock to local farmers. Global Warming Climate Change seemed to be the only shred of evidence.

Thousands of Dead Fish in Mexico

Chapter 4

Strange and Unusual 'Signs' are Occurring with the Weather

Patterns of the weather are constantly changing. News stories report that the effects of these changes are becoming a big problem all over the world. Extreme heating of the earth has caused strange and unusual storms such as volcanoes and tsunamis, earthquakes, hurricanes, tornadoes, cyclones, floods, snow, and wildfires throughout the world.

Photo showing strange activities with the weather

The heat of the earth is scorching the soil, and animals that live in the soil are dying. Because of their deaths, there is a breakdown in the ground because of the lack of oxygen flowing where these animals once lived. Thus, the deaths of the field animals are preventing the soil from being fertile. The extreme heating of the earth also has caused drought around the world, some places worse than others. The lack of water is decreasing irrigation. Thus, production of food from the ground is dropping. The demand for food and water is becoming a significant problem for the world.

Other stories state that extreme heat is a direct consequence of global warming which causes a change in the earth's climate. The change in the earth's climate causes storms such as earthquakes, hurricanes/typhoons, tornadoes, tsunamis, floods, and fires to increase in magnitude and frequency. The chapter now presents the investigation of the extreme heat and the effect it has on the storms.

Weather Sign No. 1: Unusual Temperature Changes Affecting the Weather

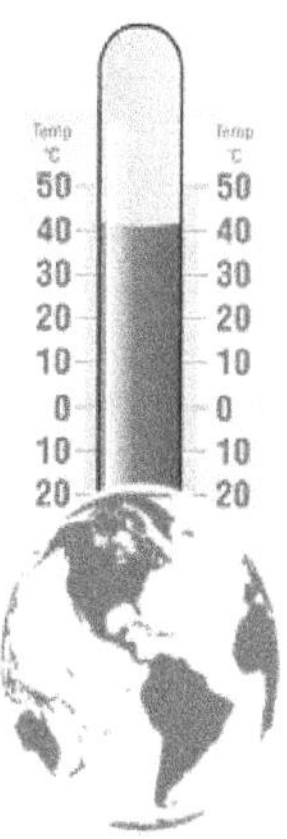

World's Temperature Increasing

First, John Vidal (2014), a writer for The Guardian, stated the year 2014 began with an unusual number of extreme weather events. Extreme weather events in (1) the equatorial and polar regions experienced extreme weather, (2) unusually heavy snowfall, (3) high monthly temperatures in the Southern Alps, and (4), unusually cold weather in the eastern United States that coincided with severe storms in Europe.

Second, the UN's World Meteorological Organization (WMO), which monitors global weather, noted that the first six weeks of 2014 saw an unusual number of extremes of heat, cold, and rain around the world at the same time. Note:

(1) Melbourne, Adelaide, and Canberra have all had record heat waves, while temperatures in Moscow were 11C above normal.
(2) Germany and Spain were 2C above normal for January.
(3) Six major depressions develop over the Atlantic.
(4) Slovenia and Australia had unusual heat waves.
(5) Snow in Vietnam and the return of the polar vortex to North America.
(6) Britain had its wettest winter in 250 years, but temperatures in parts of Russia and the Arctic were ten degrees above normal.
(7) The Southern Hemisphere had the warmest start to a year ever recorded, with millions of people sweltering in Brazilian and southern African cities (Vidal 2014).

The weather changes caused costly disruptions to transport, power systems and food production.

Third, scientists, the media, and people around the world have different views on these unusual events. A leading US meteorologist Jeff Masters stated that people are living in a time where the climate is transforming. Also, there is a reason to expect that the variations in the sea ice will have large local effects and such unexpected surprises are also because of global warming (Vidal 2014). The significant heating of the earth has brought unusual and deadly weather events. Strong storms didn't begin in 2014; they have gone on for some time. It appears, however, that the frequency and ferocity changed.

Weather Sign No. 2: Unusual Changes in Volcanoes

Volcanoes are throughout the world. Some have erupted and some are dormant. Also, there are those that are erupting. Because they are doing it on a small scale, there is not much news about such volcanoes. Noted, these are volcanoes in Africa, the Americas, Asia, Europe, Oceania, Atlantic and Pacific Oceans, and Antarctica. Also, in each of the mentioned countries and areas, together, there are hundreds of volcanoes throughout the world. These lesser-known ones only come to our attention when there is an eruption.

One place that has gotten the attention of the world is Iceland. There are approximately 130 volcanoes in Iceland, active and inactive. We can find 30 active volcanic systems under the island, in all parts of the country other than the Westfjords. On average, a volcanic eruption occurs in Iceland every four to five years. This isn't surprising considering the island has between 150 and 200 volcanoes, split into different volcanic systems. About 30 different systems are still active in Iceland, and 13 of them have erupted since the settlement of Iceland in 874 AD. Through the years, people refer to Iceland as "The Land of Fire."

One of the most outstanding volcanoes of Iceland was Eyjafjallajökull. Although it was not one of the greatest volcanoes that erupted in Iceland, the people of the world will never forget it. The entire world will remember this one.

Volcano Eyjafjallajökull of Iceland

Eyjafjallajökull's 27 March 2010 Eruption

As previously mentioned, the most notorious volcano is Eyjafjallajökull of Iceland, a volcano in southern Iceland (also known as Eyjafjöll). This volcano erupted on April 14, 2010, and lasted until October 2010. Previously, Eyjafjöll had erupted only three times in four hundred years. Its last volcanic activity was from 1821 to 1823. But the

2010 eruption of Eyjafjöll, produced ash and fire resulting in melting the glacier that covered the crater, causing massive flooding (Wikipedia).

A cloud of ash filled the sky and lasted for seven days, from April 14 to April 20, resulting in reduced visibility for pilots. The British Civil Aviation Authority ordered the country's airspace closed because of drifting clouds of ash. The entire world's communication travel system shut down. The airline companies suffered financial loss in the millions of dollars. This situation stranded many airline passengers around the world. The news media stated there was fear, confusion, and blame regarding this incident. People around the world wondered if their country would be next to be hit by an unusual event, such as this.

Volcano Grimsvotn of Iceland

In May 2011, a little over a year later, another volcano, Grimsvotn, began erupting, again causing air travel disruptions. This eruption appeared to be bigger than Eyjafjallajökull eruptions (Wikipedia). The BBC stated that the University of Iceland geophysicist Magnus Tumi Gudmundsson stated this was Grimsvotn's largest eruption in one hundred years, "much bigger and more intensive than Eyjafjallajökull." Iceland closed its main international airport and canceled domestic flights after Grimsvotn began erupting (BBC News, 2011).

Volcano Katla of Iceland

A December 2011 article posted by Jane O'Brien of the BBC News, titled "New Icelandic volcano eruption could have a global impact," predicted future devastation. There were signs that the mighty Katla volcano in Iceland, with a crater of some 6.2 miles, might erupt. There was seismic activity at this crater. The location of Katla volcano includes Iceland's largest glaciers. If the volcano erupts, there is potential for catastrophic flooding. The heat from the volcano will melt the glaciers. Billions of gallons of water could rush through Iceland's east coast and into the Atlantic Ocean. The volcano has had hundreds of tremors around the caldera since October 2011, which means that an eruption could take place any time (See photos below of flooding caused by a volcano).

Because of the health issues from the pollutants which come from the volcanoes, scientists are continuously watching those volcanoes which they classify as "sleeping dogs." If awakened, they could devastate the world and its people. The volcano that is being watched now, is the Hawaiian volcano named Kilauea.

Volcano Kilauea of Hawaii

Is this another sign of global climate change?

As of 2019, investigators found water in the base of the Kilauea Volcano in Hawaii. They feel that this finding could lead to eruptions, according to Jordan Davidson of *EchoWatch*. Last year, Kilauea, which had been erupting continuously for nearly 35 years, spewed lava that destroyed hundreds of homes on the Big Island (Allen Kim of CNN, August 6, 2019). This is the first time in recorded history that investigators discovered water on the floor of Kilauea's crater (see photo above).

The eruptions from a volcano can cause a change in the temperatures around the Earth. Science is wondering if a massive eruption will cool the earth because of the reaction of the various gases (Newsweek, 2018, Nina Godlewski). However, scientists are also cautious since the ash from volcanoes like Kilauea can have lasting effects on air, water quality, and the livelihoods of people living in the surrounding area. It was also noted that volcanoes such as this could collapse roofs, coat highways, ground airplanes, and put people at risk of developing respiratory issues. The other problem that worries scientists is the destruction of plant life by the lava. If this were to occur, it would affect the lives of animals. Another concern is famine because of the toxic sulfur dioxide from the volcano. Besides all these problems, the volcano's release of carbon dioxide will augment the effects of climate change. With the release of chlorine monoxide, there can be further destruction of the Earth's ozone layer (INVERSE, 2019).

Weather Sign No. 3: Unusual Changes in Earthquakes and Tsunamis

Earthquake damages

Deadly Earthquakes of the Past

In the past, deadly earthquakes seem to have occurred less frequently than they do now. For example, the first deadly earthquake occurred in 525 BC, and the second deadly earthquake was on October 11, 1138. Hundreds of years passed before there was the third deadly earthquake in 1556. The next deadly earthquake occurred in 1703, some 47 years later. Then, after fifty-two years, in 1755 another deadly earthquake occurred (Wikipedia).

Deadly Earthquakes Today

First, it seems that earthquakes in the twenty-first century occur more quickly and are destructive. Also, tsunamis sometimes follow earthquakes. The earthquake sometimes will create a wave, and the wave will cause the sea to rise very high. The water is so high and moving at such a great speed that it washes over the land at the same rate. The devastation from a tsunami is enormous. It submerges people, livestock, houses, farmlands, cars, and other utilities on the lower areas of the land where it comes ashore.

Tsunami Wave

Second, the news media shares the story of the deadly earthquake which set off a tsunami on December 26, 2004. The earthquake, with a magnitude of 9.1, took place in the Indian Ocean and triggered a tsunami that caused a death toll of 230,000 people. The news sources stated that this earthquake caused the entire earth to vibrate four-tenths of an inch and caused other earthquakes as far away as Alaska. The effects of this unusual event affected fourteen countries.

National Geographic magazine compared the tsunami to the bombing of the island of Hiroshima. The devastation of the tsunami, however, was much worse—like dropping a bomb that was twenty-three thousand times stronger than the bomb dropped on Hiroshima. It killed more people than all the earthquakes that took place between 1755 and

1998 combined. In that 243-year period, 136,981 people died, but this event killed 230,000 people in one day, almost double the number of individuals killed in 243 years by earthquakes/tsunamis (Wikipedia).

Third, six years later, the news media story states how on January 12, 2010, in Haiti, an earthquake with a magnitude of 7.0 killed approximately 222,570 people. This earthquake destroyed the city of Port-au-Prince. Apart from the thousands who died, it wounded thousands more of the inhabitants, and left tens of thousands homeless. The lack of supplies for the victims caused diseases to increase among the remaining population. Support from around the world came to assist the people of Haiti. To this day, Haiti and its inhabitants are still trying to recover from this disaster. Prior to the 2010 earthquake, there hadn't been an earthquake like this one in the West Indies for over a hundred years (Wikipedia).

Fourth, a few weeks after the earthquake in Haiti, another news media story informs the people about the deadly earthquake that took place in Chile, South America, on February 27, 2010. The earthquake in Chile was a magnitude of 8.8 and lasted for three minutes. It was so powerful that scientists noted it shortened the length of the day by 1.26 microseconds and moved the earth's axis of rotation by three inches or 2.7 milliarcseconds. As far north as the city of Ica in southern Peru, people felt the shaking of the earth. Fifty-three countries issued tsunami warnings. This earthquake destroyed half a million homes, and about 723 people died (Wikipedia).

Fifth, there are many stories about earthquakes. However, since 2007, hundreds of earthquakes have taken place in the world. Places such as the Ionian sea, Italy, Chile, Alaska, California, Switzerland, Turkey, Kermadec Islands, Indonesia, Mexico, Hawaii, Xizang, France, Canada, Albania, Argentina, Italy, Peru, Haiti, Pyrenees, Bouvet Island, Virgin Islands, Banda Sea, Tonga, Mariana Islands, Iran, Greece, Dominican Republic, Philippines, Japan, Guatemala, Australia, Costa Rica, China, many areas in the United States (Idaho, Oklahoma,), and the list goes on where areas have experienced earthquake activity.

The earthquakes cause much damage and most times, the lives of people are a loss. For example, on September 30, 2009, a series of earthquakes occurred in Sumatra in Indonesia and Tonga, where it left 600,000 people homeless, and approximately two hundred died because of flooding (Liz O'Neill and Michael Hill). Just imagine the damage the earthquakes cause to countries and their people.

Sixth, as recent as December 28, 2019, to January 2020, a series of earthquakes struck the island of Puerto Rico, including six that were magnitude 5 or greater. Much damage has occurred. On January 13, the news media reported that the earthquakes have destroyed or damaged 3,000 homes. Three people died and they calculated financial losses at $3.1 billion US dollars (Wikipedia). Note the photos below.

Photos: Structural damages by one of Puerto Rico's Earthquakes

Photos: Structural damages by one of **Puerto Rico's** Earthquakes

Seventh, World News stated on January 25, 2020, a 6.7 earthquake struck Turkey near the town of Service in eastern Elazig province on Friday evening. The earthquake caused the downing of at least 10 buildings and killed 22 people and wounded approximately 1,243.

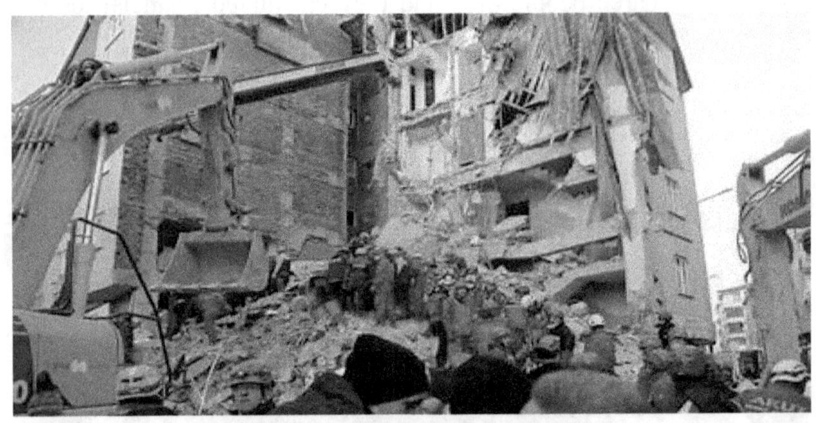

Photo: Turkey's Earthquake Destruction

Weather Sign No. 4: Unusual Changes in Hurricanes

Hurricanes Increasing in Frequency and Ferocity

Hurricanes are increasing in frequency and with greater devastation, as are typhoons and tornadoes.

First, in 2005, of the seven major hurricanes— Dennis, Emily, Katrina, Rita, and Wilma—were deadly and caused much destruction. Hurricane Katrina laid waste to New Orleans. Many people died and it destroyed many homes. The levies could not hold the pressure of the water, and the water broke through and flooded the city. At least 1,836 people died in the hurricane and in the subsequent floods, making it the deadliest US hurricane (Wikipedia).

Second, the year 2005 was a year of extremes for hurricanes. The extremely hot weather formed twenty-eight tropical storms and fifteen hurricanes. The most intense hurricane formed in the North Atlantic Basin, based on barometric pressure, was Hurricane Wilma. The most damaging was Hurricane Katrina, which caused $81.2 billion in damages (Blake and colleagues 2007).

Third, on November 8, 2009, Hurricane Ida took the lives of ninety-one persons in Mexico and caused great flooding up the Eastern Seaboard of the United States. Also, the hurricane's destruction to the power system caused thousands of people to be without power in their homes and businesses.

Fourth, on October 23, 2015, Hurricane Patricia, another powerful storm on the West Coast of North America was the second-most intense tropical cyclone on record worldwide. It had maximum sustained winds of 215 miles per hour. It made landfall, with winds of 150 miles per hour. The hurricane produced widespread flooding in Central America. The hurricane affected hundreds of thousands of people in Guatemala. Also, it killed six people; four in El Salvador, one in Guatemala, and one in Nicaragua. Torrential rains extended into southeastern Mexico. The authorities calculated the damages to be in the millions (Wikipedia). The storm shows that hurricanes are increasing in strength.

Fifth, on September 1, 2019, Hurricane Dorian struck the Abaco Islands as a Category 5 hurricane. A day later, on September 2, 2019, it hit Grand Bahama Island in the same category. The unfortunate effect on the island, the hurricane then stalled over Grand Bahama for another day, finally pulling away from the island on September 3. Hurricane Dorian, a monster storm struck the Bahamas with punishing 185 mph sustained winds and 25 feet storm surges.

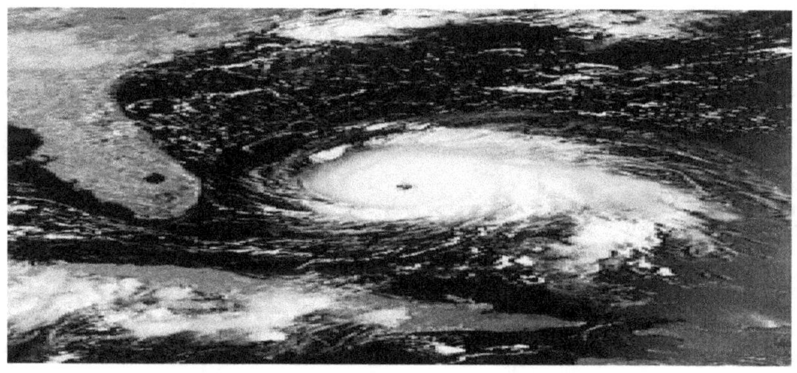

Category 5 Hurricane Dorian over the Bahamas

According to a new report by the Inter-American Development Bank, the hurricane's damages on these two Bahama islands caused an estimate of $3.4 billion. The damages put the Bahamas on a difficult path to reconstruction since it is equivalent to one-fourth of the country's gross domestic product. That's equivalent to the United States losing the combined economies of Florida, California, and Texas, the report said. Also, it left 29,500 people homeless or without jobs — or both (The Miami Herald, 2019). After a full evaluation of Hurricane Dorian's damages, the authorities of the Bahamas estimated damage was over US$7 billion, and at least 65 deaths in the country. As of that date, the rescue workers could not find 282 people (Wikipedia).

View some of the destruction on the Bahamas Islands of Great Abaco, Grand Bahama, and Turtle Cay, as depicted in the following photos:

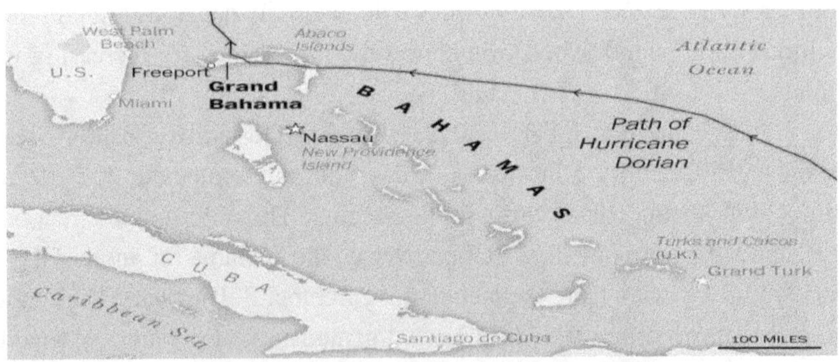

Category 5 Hurricane Dorian's Path – 2019

Hurricane Dorian Destruction of Great Abaco Island, Bahamas 9-1-3, 2019

Hurricane Dorian and its effects on Green Turtle Cay

Hurricane Dorian's Destruction on the Grand Bahama Island

Climate Change is Increasing the Frequency and Ferocity of Hurricanes

The New York Times reports that "the effect of the climate crisis is increasingly clear. The warmer oceans intensify the storms' force. So, these become more powerful and more dangerous. A warmer atmosphere and higher sea levels mean storms can hold significantly more moisture, making wetter storms. Heavier storms stall and drop enormous amounts of water. This happened with Hurricane Harvey that hit Houston, Texas and dropped 60 inches of rain. Also, the most recent was Hurricane Dorian that stalled over Great Abaco and Grand Bahama Island. The results of such stalling caused thousands of homes to be destroyed and the deaths of 65 people. The news media labeled the behavior of Hurricane Dorian as the result of global Climate Change. This hurricane has given much news about global climate change, especially in the United States of America. Both Islands of the Bahamas, Grand Bahama, and Great Abaco have sustained extensive damage, loss of lives, and remain underwater, as of September 3, 2019.

Weather Sign No. 5: Unusual Changes in Typhoons/Cyclones

Similar to hurricanes, typhoons and cyclones are also very destructive. On September 24, 2009, Typhoons Parma and Ketsana produced massive amounts of rain. The same amount of rain that usually falls in one month in this area fell in just twelve hours, and three hundred people died. On February 27, 2010, a storm named Xynthia went ashore in Europe. Clouds stretched from the Atlantic Ocean to northern Italy. This powerful storm brought hurricane-force wind gusts, flooding rains, and a three-foot storm surge, topped by thirty-two-foot battering waves that overwhelmed the sea walls in France, killing scores of people (NASA–Earth Observatory 2010). The authorities estimated damage from the storm at $1.5–3 billion (Air Worldwide 2010).

On November 8, 2013, Typhoon Haiyan, known as Typhoon Yolanda, devastated portions of Southeast Asia, particularly the Philippines. This typhoon was one of the strongest tropical cyclones ever recorded and the deadliest Philippine typhoon. At least 6,268 people died in that country alone, and it left many people homeless. It affected about eleven million people (USAID 2013). About eleven million people were affected, many of whom were left homeless (USAID 2013).

Weather Sign No. 6: Unusual Changes in Tornadoes

Tornadoes affect the world in areas such as America, Canada, Russia, Bangladesh, South Africa, Australia, New Zealand, Great Britain, and Argentina. New data indicate that the central states of the USA have the highest tornado risk in the world. In 2011 the United States experienced a truly bizarre tornado season. Approximately six hundred tornadoes took place across America in one month, the largest number of storms in a single month inside the United States (Stormtrack 2007).

Photo showing the power of Tornadoes

The average number of tornadoes in the United States is about 1,200 per year, but in 2011, the massive tornado outbreak in the Southeast at the end of April was the worst natural disaster since Hurricane Katrina. One tornado that went through Tuscaloosa, Alabama, was a mile wide, and winds estimated at 260 miles an hour. Also, the tornado that went through Joplin, Missouri, in 2011 was the deadliest tornado in over sixty years (Snyder 2011).

By the end of June 2014, 625 tornadoes had already occurred (Hoppe 2013). The Daily Mail Online reviewed photos of a land area before and after a tornado onset and concluded, "After viewing pictures such as those, what else is there to say?" The weather patterns have changed and are continuing to change. Why? (Stevens 2011). The answer, global climate change is causing changes in the temperature of the earth and thereby, causing changes in the weather.

Weather Sign No. 7: Unusual Changes in Flooding

Flooding has been devastating in recent years, causing billions of dollars in losses, and it appears to be increasing around the world. Around 47 percent of overall losses were from inland flooding in Europe,

Canada, Asia, Australia, the United States, and many more countries (Hoppe 2013).

Photo showing a Flooding Process

Flooding in Germany

Flooding in the first part of 2013 was the most expensive drain on Germany's economy. The frequency of flood events in Germany and central Europe has doubled since 1980. Flooding is now an increased hazard. The danger from flooding has become a burden on European governments as they must get better flood control and update risk awareness (Hoppe 2013).

Flooding in Australia

Australia went through a ten-year drought, but from the end of 2010 to 2013, devastating floods came upon the land and people without mercy. As the years went by, the floods in Australia increased. For example, from 1806 to 1909, there were eight notable floods. From 1909 to 2010, there were twenty-nine significant floods. From 2010 to 2015, however, there were ten notable floods. Flooding in Australia is increasing in frequency. The scientific community has blamed global warming for the flooding (Wikipedia, Australian Flooding). Also, in the years following 2015, there were other significant floods.

In 2016 there was an increase in the rains that flooded central New South Wales, western Victoria, parts of western Queensland, and areas around Adelaide. In 2017, floodwaters stranded tens of thousands of Australians. Also, in 2018, the rains of February, March, November, and December, caused devastating flooding.

In the year 2019, the months of February, August, and September yielded much flooding. The flooding caused the loss of the lives of people and animals. However, although several homes and property sustained damages, the rescue efforts saved many people from drowning. The authorities believe that climate change is creating these flooding episodes in Australia. Not only is Climate Change causing massive flooding in parts of Australia, but in other parts, because of the heat and drought, as of 2019 Australia is burning with wildfires.

Flooding in Canada

Starting in 1879, the flooding events were not as frequent. However, as the years went by, the frequency changed. The history of floods are: 1879 Fort Calgary; 1894 Fraser River flood; 1897 Fort Calgary; 1902 Saint John River; 1915 Calgary; 1929 Calgary and southern Alberta flood; 1929 Tsunami Burin Peninsula, Newfoundland; 1948 Fraser River flood; 1950 Red River flood; 1954 Toronto region flood: Hurricane Hazel; 1973 Saint John River flood; 1974 Grand River flood; 1979 Tropical Storm David's flooding of Moncton; 1984 Pemberton Valley flooding; 1986 Winisk flood; 1987 Montreal Flood; 1996 Saguenay flood; 1997 Red River flood, "flood of the century": a return interval ranging from 100 to 500 years; 2003 Pemberton— sea to sky flooding; 2004 Alberta; 2005 southern Alberta; 2007 Saskatchewan; 2008 Saint John River flood; 2009 Red River flood; 2010 Southern Alberta and Saskatchewan flood; 2010 Hurricane Igor in Newfoundland; 2011 Assiniboine River flood; 2012 Thunder Bay to Montreal; 2013 Calgary and Southern Alberta flood; 2013 Southern Ontario flash flood (Wikipedia)

In 2014, there were the Manitoba and Alberta floods (bing.com). In 2015 there were floods in Canada again. The Huffington Post reported in an article "Climate Change: Floods in Canada Are Going to Get

Uglier," that people were identifying the change in the weather with global warming.

In 2015, Flooding destroyed many properties, and the government issued several state of emergencies. Another article showed, "Floods are by far responsible for most of the natural disasters in Canada.

In 2016, the flooding continued, and there were many. Moisture remaining from Hurricane Matthew fueled a weather system in Canada that unleashed heavy rains in parts of the eastern province of Newfoundland, resulting in significant flooding and damages (Wright 2016); (CBC News 2016).

PR Newswire (2017) noted people's great concern about freshwater, as flooding contaminates the drinking water. In 2018 and 2019, Canada experienced much flooding along with other snowstorms and drastic low temperatures. The authorities blamed Climate Change as the cause of the strange and unusual events.

Flooding in the United States of America

In the United States, the areas affected by flooding were Atlanta, Tennessee, Arkansas, northern Mississippi, Missouri, and southern Kentucky. In Atlanta in 2010, ten people died because of flooding. Also, in May 2010, Tennessee's rainfall for two days was greater than 19 inches. As a result, the Cumberland River in Nashville rose to 51.86 feet, a level not seen since 1937. Floods affected the area for several days afterward and caused the death of twenty-one persons in Tennessee (Wikipedia).

Over five inches of rain fell in the Little Rock area of Arkansas and over ten inches in northeastern Arkansas, closer to the Mississippi River. Also, up to eight inches of rain fell in West Memphis, and similar amounts were recorded across western and southern Kentucky, where over seven inches fell in the Hopkinsville area and up to four inches across the Missouri bootheel. Besides the torrential rains, moist air and instability contributed to multiple tornadoes in the same areas, killing four people in Mississippi and one in Arkansas.

During 2015 several coastal states experienced severe flooding. Heavy rainfall and high-water levels combined to cause compound flooding, which resulted in increased flood risk for many major US cities in the last century (Carbon Brief in USA 2015). In June 2015, deadly floods took place in the central United States. In parts of Oklahoma, Texas, and Louisiana, floods caused the deaths of two individuals. A mother and two of her children died in flash floods in Ohio in July 2015, when the floods swept away the family's mobile home (floodList 2015).

Shreveport, Louisiana, and the parishes of Bossier and Caddo experienced significant flooding after the Red River overflowed. For this area, this was unusual, since there had not been such flooding of this sort in fifty years. President Obama declared the flooded area of the Red River in Louisiana a disaster area from May 18 to June 20, 2015. Also, in July, the government declared a state of emergency because of the flooding in Kentucky. The authorities also stated that two people died, and several other people were missing.

The cost of flooding in Texas and Oklahoma was $3 billion. In this flooding, people lost their lives. For Texas and Oklahoma, it was "historic flooding." In an article titled "Model Helps City Planners Prepared to Weather Floods and Storms," it stated that climate change may be a global-scale debate, but the University of Wisconsin–Madison researchers know that preparing for climate change's impact on weather is a profoundly local problem (floodList, Carbon Brief in the USA). The article warned the leadership that climate change is real, and they need to prepare for these types of storms.

Catastrophic flooding occurred throughout South Carolina. The level of water in the roads from flooding rivers had reached their highest levels in decades. The National Weather Service reported that the state received nearly seventeen inches of rain in seventeen hours, and there was still more rain to come. The state experienced so much rain that the authorities had to warn residents to stay off the roads because the conditions of the roads were "changing by the minute" (Manzo & Colleagues 2015). CNN stated that after the rain stopped in South Carolina, the state had to grapple with a host of new concerns. Dams broke, and there was damage in the billions of dollars. Governor Nikki Haley called the deadly rain a "thousand-year event" but that "God smiled on South Carolina because the sun is out" (CNN).

Flooding in Brazil

Rain flooded the country of Brazil in 2011. It was the country's worst flood disaster on record. CNN stated that the death toll reached 655 in the mountainous region northeast of Rio de Janeiro (De Moura 2011). Teresopolis, near Rio de Janeiro, was the area hardest hit by the flooding. Brazilians were expecting more rain and more potential landslides. There were 730 confirmed dead and at least 207 people missing. The authorities feared that the death toll could reach one thousand (*PI News* 2011). The Wall Street Journal reported that despite a massive recovery effort, it was tough to get an accurate count of victims. In many areas, because of the mudslides, entire families were unaccounted for. Landslides destroyed whole communities in some areas; individuals died, and the continuous rain and flooding left nearly fourteen thousand homeless (Prada & Kinch 2011).

Flooding in other Areas of the World

On September 30, 2009, rain in Samoa flooded southern India and caused 2.5 million people to become homeless. The area had not had this amount of rain in sixty years (O'Neill, 2009).

The flooding that took place in 2010 in Iceland was very unusual. During the eruption of the volcano, fire, smoke, hot rocks, and lava came out of its crater. The heat from this explosion caused the surrounding sheets of ice to melt. This event caused massive flooding. Eight hundred people had to flee for their lives (Than 2010). Eight thousand families had to evacuate in Asuncion, Paraguay after the Paraguay River overflowed (Davies 2015). In Guinea after nearly eighteen inches of rain in three days, flooding in Conakry left people dead. Flooding in Odisha, India, which was caused by the overflowing of three rivers, resulted in deaths, and thousands of people were affected. In western India in the areas of Madhya Pradesh, Rajasthan, and Gujarat, some twenty-eight persons died (The Indian Express 2014). In North Vietnam, twenty-eight people died because of flooding and landslides (Davies 2015). In Pakistan's flooding about sixty-nine people were killed (Shai 2014).

There are so many more cases that detail the devastation caused by rainfall and flooding with the destruction of property, losing lives, and the financial burden floods have on the affected countries. People all around the world share the same sentiment why these tragedies are taking place and, sometimes, more rapidly than ever—they want answers. Science says its Global Warming and Climate Change. But is there another reason why these strange and unusual events or Signs are taking place in the sky, on the land, in the waters, and with the weather?

Weather Sign No. 8: Unusual Fires

Science says that the climate is transforming. Extra heat is like an extra shot of gasoline for forest fires. The increased evaporation caused by excess heat makes wood in the forests drier. If a fire starts, the high winds fan the fire and create a blaze which sometimes results in a firestorm, causing much damage and even loss of lives.

Photo showing 'Forest Fires'

A **firestorm** is a conflagration that attains such intensity that it creates and sustains its own wind system. It is most commonly a natural phenomenon, created during some of the largest bushfires and wildfires (Wikipedia). Many firestorms occur in the forested areas of the United States and Canada. Also, we find them in the vegetated areas of Australia and South Africa. It only takes a spark from lightning or human ignition to start a firestorm., as seen recently in Queensland, Australia, and California, USA.

NASA's Research on Firestorms

First, in October 2011, NASA released a series of new satellite data visualizations that showed tens of millions of fires detected worldwide from space since 2002 (NASA.gov). Laura Betz, another scientist at Goddard Space Flight Center, noted that in June 2012, fires burning in Siberia sent smoke across the Pacific Ocean and into the United States and Canada (Seftor & Betz, NASA.gov).

Second, burning fires emit carbon dioxide. This and other greenhouse gasses moving among the land, air, and ocean affect climate change. Scientists agree that the earth is the "fire planet." Also, "On an average day in August, NASA's instruments detect some 10,000 actively burning fires around the world."

Third, scientists concluded that fire affects 30 percent of the earth's land surface. Louis Giglio, another University of Maryland scientist, revealed that they are "still trying to determine how fire patterns are changing as global temperatures rise and precipitation changes." However, they concluded that people of the earth are in for a smoky, diseased, and fiery life if they don't make changes quickly (NASA.gov).

Firestorms of the World

The news media is constantly presenting stories about the firestorms of California, the United States, Queensland, Australia, the Amazon, South America, and those in Africa. Remember, wherever there are firestorms, the world is being affected, as seen in NASA's research.

California Firestorms

The firestorms of California carried headlines such as (1) Hell Hath No Fury Like Raging California Wildfires (2017), (2) California's Burning: The Social and Political Background of Deadly Infernos (2018), (3) Welcome to the New World of Wildfires (2017), (4) Global Fires and Droughts: The Media Cover-up of Climate Change (2018), (5)

Raging California Wildfires (2018), (6) California Wild-Fires: Paradise Lost, The Town of Paradise in Northern California Destroyed (2018

The Amazon's Rainforest

The firestorms of the Amazon's Rainforest had headlines such as, (1) "What you need to know about the Amazon rainforest fires," (2) "What would the Earth be like without the Amazon rainforest?" (3) "Amazon rainforest fires: What's happening now and how you can help," (4) "5 highly-ranked charities helping with the Amazon rainforest fires," (6) "The Amazon rainforest is on fire. Climate scientists fear a tipping point is near," (7) "Brazil's President: We don't Have The Resources To Fight Amazon Rainforest Fires," (8) "How to Help the Amazon Rainforest As Brazil's Wildfires Ravage the 'Lungs of the Earth'," (9) "Putting out the Amazon fires isn't just a physical challenge – it's a political one," (10) "The Amazon rainforest is on fire. The president of Brazil is defiant, he suggested that people who dislike him set the fires, (11) "G-7 nations pledge $40 million to fight Amazon fires," (12) "Africa: Opinion – Unfortunately, the Amazon Isn't the Only Forest Fire Story."

The above are a few of the 190 stories listed on just one website. Each headline tells a story of the burning firestorm. Some of these stories were political, some have brought attention to those who care about various other parts of the world where fires are burning, say Africa. Also, there were stories that implied the news was fake. However, they were all stories about Global Climate Change and its effects.

The Effects of Firestorms and Climate Change

Firestorms affect the earth, humans, animals, and trees. The effect of the fires affects the oxygen supplies for humans and animals, and the CO2 levels for the trees. However, if the trees are being cut down from the earth the CO2 levels increase, thus, increasing the effect of climate change.

Weather Sign No. 9: Unusual Snowstorms

There have been changes in the snow patterns, locations, and the presentation of snowstorms. The frequency of extreme snowstorms in the eastern two-thirds of the contiguous United States has increased over the past century (National Centers for Environmental information – NOAA). The many snow blizzards in the Northeast through the years have affected property and people. This chapter presents some of these blizzards and the damages they caused.

Snow Blizzards 1888 – 1900's

The 1888 blizzard dumped over 50 inches of snow. The February 1899 blizzard dumped 20 inches of snow. The 1913 blizzard killed over 250 people and left some 35 feet of snow. The 1922 blizzard killed 98 people and injured about 133 others. The November 1940 storm killed 145 people and the November 1950 storm killed 353 people. The Blizzard of 1967 killed 67 people. A January 1975 snowstorm's heavy snow and cold killed over 100,000 animals. Also, about 377 people lost their lives. The March 1993 storm was both a blizzard and a cyclone. The storm was responsible for 310 deaths and $6.6 billion in damage, and it shut down the South for three days.

Snow Blizzards 2000 – 2016

The 2000 blizzard was 150 miles wide and 500 miles long. It did great damage. Two blizzards in February 2010 (Snowmageddon) broke snowfall records in the mid-Atlantic region with an accumulation of 32.4 inches of snow at Washington's Dulles International Airport.

A 2011 blizzard dumped over 21 inches of snow in Chicago.

The 2012 blizzard not only dumped snow but also produced tornado warnings and watches to the central and southern United States.

The 2013 blizzard had hurricane-force winds. Some 650,000 homes and businesses lost power and five people died.

The 2014 blizzard, known as "Snow-vember," deposited five to seven feet of snow. At least fourteen people died. Snow remained up to eight months after the storm.

The 2015 blizzard had heavy snow mixed with winds over 35 miles per hour. Snow totals reached up to 30 inches within two days. US

President Obama made a federal disaster declaration for the snowstorm, allowing some reimbursement for damages.

A 2016 blizzard produced up to three feet of snow. The news media nicknamed this historical storm "Snowzilla."

Recent Research on Snowstorms

First, recent research has shown that Global Warming/Climate Change is increasing surface temperatures and reducing the Arctic ice. The result will increase winter storms in the eastern United States. Also, high pressure blocking patterns over the Atlantic will continue to produce cold outbreaks and together with slower-moving weather systems will make the persistence and severity of winter storms worse (AGU100 – Advancing Earth and Space Science (Francis & Vavrus, 2012).

Second, according to another scientific research, one outcome shows that in response to global warming and climate change, it projects that "the frequency and intensity of extreme weather events will increase (Schwartz and Schmidlin, 2002).

Third, close observation shows that the weather-events appear to be the most visible strange and unusual signs to humans. Could anyone imagine snowstorms in Hawaii in July? But it happened (Fox, C. (2015).

Closing Chapter Comments

The News Media, their stories are informing people about the strange and unusual 'signs' that have and are occurring on Planet Earth. The news media also stated that science is trying to find answers for the reasons (1) birds are falling from the sky dead, (2) Cyber warfare is being conducted in the sky, (3) the occurring meteorite incidents, (4) mass deaths of animals on the land, (5) increased heating of the land causing unusual storms, and the serious effects. People are seeking answers about these strange events. Do you think Science has the answers for such strange and unusual signs? Well, guess what's in the following chapter? Let's read!

SECTION TWO

THE SCIENCE OF

GLOBAL WARMING - CLIMATE CHANGE

Every good and perfect gift is from above,
coming down from the Father of the heavenly lights,
with whom there is no change or shifting shadow. ***(James 1:17)***

But to each one of us, grace was given
according to the measure of Christ's gift.
Therefore, He says: "When He ascended on high,
He led captivity captive and gave gifts to men. ***(Ephesians 4:7-8)***

If any of you lacks wisdom, you should ask God,
who gives generously to all without finding fault,
and it will be given to you. ***(James 1:5)***

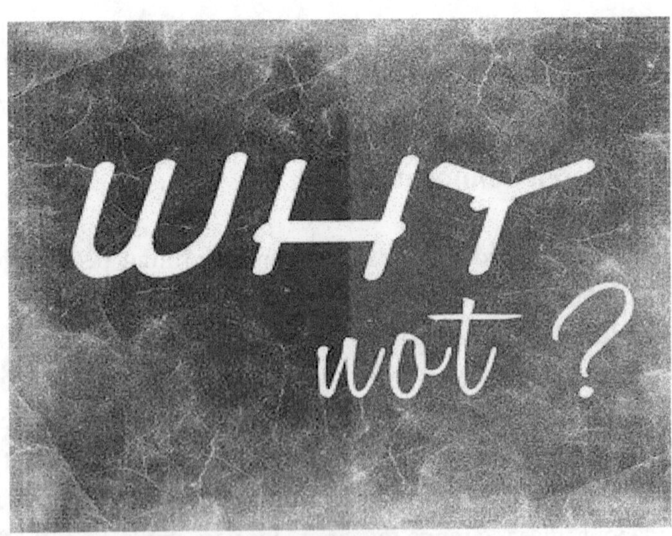

Science Research

Chapter 5

SCIENCE ATTEMPTS TO EXPLAIN THE STRANGE AND UNUSUAL 'SIGNS'

Science is Key

Everyone seems to want answers regarding the strange and unusual 'signs' that are occurring in the sky, on land, in the waters, and with the weather. The strange and unusual signs are increasing in frequency and with greater powers of performance. The news media stated that people of the world are looking to scientists for answers. Science is continuing to investigate the strange and unusual signs which are occurring in the world. Maybe new scientific findings could provide answers for these unusual signs that are being shared by the news media.

People Seek Answers for New Media Unusual Stories

First, the media stories of strange and unusual events that are taking place in the sky. Note, (1) Birds falling from the sky dead; (2)

Meteorites are falling from the Sky to the earth; and (3), Cyber Warfare Rages Across the Sky.

Second, the news media stories about strange and unusual signs on the land. Note, (1) Unusual Activities among Dictators; (2) Hate display among Humankind; (3) Unusual Activities among People of the Earth; (4) Increased Incidents of Uprisings, Violence, Terrorism, and Wars; (5) Modern-Day Pirating; (6) Worldwide Terrorism; (7) Fleeing Refugees; (8) Unusual Plagues and Diseases; (9) Unusual Instability in Leadership and Government; (10) Unusual Signs of the Failing World Economic Systems; (11) Unusual Employment Failures around the World; and (12) Unusual Activities among Animals.

Third, the news media stories about strange and unusual signs in the Waters. The stories stated that Freshwater and Sea Animals are dying by the thousands; and there seem to be no clear answers for their mysterious, unusual, unbelievable, or phenomenal deaths.

Fourth, the news media stories about the strange and unusual signs about the weather on earth. The stories about the weather appeared to be the main foundation for all the other strange and unusual signs that are occurring.

The chapter tries to put together all this relevant information in the book. The purpose is to allow readers to gain the needed knowledge that they may understand the truth of the strange and unusual events that are taking place on the earth. Therefore, at this point in the book, the chapter gives answers by the scientific community for the signs /events in the sky, on the land, in the waters, and with the weather. Let us now continue to read!

Science Explanation for Unusual Events in the Sky

As previously mentioned, the news media reported on strange and unusual events taking place in the sky, most notably the thousands of birds that fell from the sky, dead. The scientific community offered various answers, including that they were "unsure of the cause of mass death of birds in Davis County." They did not know why the birds died.

Scientists Gave Ten Theories for the Strange and Unusual Deaths of The Birds and Fish

At the beginning of the questioning by the news media and laypeople, the answers the scientific community gave appeared vague. There are ten theories for the strange deaths of the animals, especially the birds. The theories are (1) mainstream explanations, (2) meteor showers, (3) New Madrid fault line, (4) government testing, (5) GMO mutation, (6) geoengineering, (7) HAARP, (8) scalar weapons, (9) Project Blue Beam, and (10), geomagnetic and other earth changes (West & Gardner – Activist Post, 2011).

Each theory offers a scientific answer for the death of thousands of birds and fish. As readers view these theories, remember that a theory is educated speculation, not actual fact.

The Scientific Theories for the Animals' Deaths Explained

First, the *mainstream claims* that lightning, hail, mid-air collision, power lines, and even fireworks killed the birds. Since there were no roasted birds, the news media stories labeled the claim as "ridiculous" because "Birds are incredibly sensitive to their environment (West & Gardner, 2011).

Second, the *meteor showers theory* claims that the explosions and shock waves from the meteor showers killed the birds.

Third, the *New Madrid fault line theory* explains that we could relate the deaths to earthquake activity because of a fault line that runs along the Midwestern section of the United States (Mercola 2011).

Fourth, the *government testing theory* – maybe various US government testing with bioweapons may have killed the birds (and fish) (Smith, 2012).

Fifth, the *GMO mutation theory* presented evidence that die offs are happening across species. Just as in the bee and bat populations, the same happened with the birds and fish.

Sixth, the *geoengineering theory* claimed that spraying of chemicals may have something to do with the deaths of the birds and fish.

Seventh, the *High-Frequency Active Auroral Research Program* (HAARP) gives off secret electromagnetic warfare capabilities in the form of frequencies. These frequencies may have short-circuited the navigational systems of the birds and caused their deaths.

Eighth, the scalar weapons theory states that scalar technology which has the potential to turn the environment itself into a weapon and a merciless killer may have killed the birds flying within the range of its operation.

Ninth, the *Project Blue Beam theory* uses the natural energy present in the ionosphere as both a visual and an acoustical device. This theory argues that the Project caused some unintentional misstep that caused death to birds and fish.

Tenth, the *geomagnetic and other Earth changes theory* state that tampering with the universe is causing many unusual and strange activities to take place. For example, the dwindling magnetosphere and falling oxygen levels which this system produces could cause creatures such as birds and fish to die because of the lack of oxygen (German 2013). The other concern about the geomagnetic and other earth changes, the increase in the sun activity results with magnetic storms. These magnetic storms may eventually weaken our overall natural habitat (Tyson 2003).

However, as science advanced over the years, the research has shown that the real cause of the strange and unusual events is because of Global Warming and Climate Change. It is a scientific belief that not only are the deaths of the animals in the sky, but also the deaths of the animals on the land and in the waters, Global Warming/Climate Change affected. Which, per the science, the real cause is human behavior - their treatment of the planet.

Science Explanation for Unusual Events on Land

Because leaders and governments cannot give answers, for the strange and unusual events on the lands of the earth, people are looking to the scientists for answers. The following are twelve areas of concern: (1) *Unusual Activities among Dictators;* (2) *Hate display among Humankind;* (3) *Unusual Activities among People of the Earth;* (4) *Increased Incidents of Uprisings, Violence, Terrorism, and Wars; (5) Modern-Day Pirating;* (6) *Worldwide Terrorism;* (7) *Fleeing Refugees;* (8) *Unusual Plagues and Diseases;* (9) *Unusual Instability in Leadership and Government;* (10) *Unusual Signs of the Failing World Economic Systems;* (11) *Unusual Employment Failures around the World; and* (12) *Unusual Activities among Animals.*

The research about these 12 areas of concern is all wrapped up in the package of "Human Behavior." Note some of the selected science reports that follow.

Science Explains Humankind's Behavior in the World

Studies have shown that violence is found throughout recorded human history, leading some researchers to conclude that we crave it, that it's in our genes and affects reward centers in our brains. However, going back millions of years, evidence suggests our ancient human ancestors were more peace-loving than people today, though there are signs of cannibalism among the earliest prehistory humans. "Humans certainly rank among the most violent of species" (LiveScience 2006).

Besides violence, the behavior of "theft" is in our genes. According to other studies, they say humans 'steal and cheat.' The worst cheaters are those with high morals who also, in some twisted way, consider cheating to be an ethically justifiable behavior in certain situations.

The findings of Jardine, states that (1) the human behavior is an innate human defiance, (2) it has a need for social acceptance, (3) it has the inability to understand the nature of risk, (4) has an individualistic view of the world and the ability to rationalize unhealthily, and (5), has a genetic predisposition to addiction. People justify bad habits, she states, by noting exceptions to known statistics, such as "It hasn't hurt me yet," or "My grandmother smoked all her life and lived to be 90" (Bryner 2008).

Other studies agreed with the above expressions of human behavior as being accurate. These kinds of actions often lead to unrest, turmoil, and wars. Wikipedia states that "war is generally characterized by extreme violence, social disruption and an attempt at economic destruction. It can be defined as a form of political violence or intervention."

Based on the above underpinnings about 'human behavior,' the former twelve areas of concern can rest. So, as humans interact with each other and their environment they form behaviors. As a result, dictators arise, hate grows, uprisings, violence, terrorism, and wars. From these come fleeing refugees, unusual plagues, and diseases. The outcomes most times are too great for the world leaders. Therefore, people observe a yield of poor leadership, instability in governments, and failing economic systems. However, leaders try to improve their economy by increasing production. But science states that from these efforts to produce, pollutants and waste results in the earth's warming and climate change.

The climate change phenomenon is worth a scientific explanation so that people can understand how human behavior has added to the earth's warming and the climate change phenomenon. Thus, producing many other strange and unusual signs or events on our Planet Earth. The findings may satisfy the curiosity that people of the world are trying to satisfy.

Remember now, people and science have debunked the old theories about the cause of the strange and unusual events/signs. New findings have identified Global Warming and Climate Change as the reason for these signs that are taking place on Planet Earth. This belief has now gone worldwide. However, some people accept the science of Global Warming/Climate Change, but others deny it. Some people even said, "science is 'fake news.'" Then, there are those people whose belief is that God is in control, and He is getting the attention of people for these signs are a reminder of the 'End Time.' So, who is right? Ok, let us investigate the information scientists are presenting about Climate Change and see what it is all about? Readers may find interesting information.

Chapter 6

SCIENCE SAYS GLOBAL WARMING CLIMATE CHANGE IS CAUSING UNUSUAL SIGNS ON PLANET EARTH

Note the Changing Climate

It is a fact that strange and unusual events which have been identified as 'Signs' are occurring all over Planet Earth. Science claims that global warming/climate change is the contributing factor for unusual and strange activities or signs. So, science now tries to explain why these events or signs are occurring.

To better understand science's phenomenon of global warming/climate change and how it affects the world, four areas discussed which explains the science. These are: (1) The definition of Global Warming and Climate Change; (2) Various Ideas about the Phenomenon of Global Warming – Climate Change; (3) Science's

Reality of Global Warming – Climate Change is Real; (4) The United States' Intervention in Global Warming – Climate Change Prevention Effort, Indicates that the Science is Real. Let us investigate!

Global Warming and Climate Change Defined

Global Warming is a term used to describe the current increase in the earth's average temperature. Climate change refers not only to global changes in temperature but also to changes in wind, precipitation, and the length of seasons, as well as the strength and frequency of extreme weather events like droughts and floods. The terms *global warming* and *climate change* are often used interchangeably in newspapers and television reporting. However, these are two distinct systems, but each system assists the other, and together outcomes are generated. For example, global warming refers to the rise in the global average temperature of the earth. When the temperature of the earth rises and warms the earth, it causes a change in the earth's climate.

Climate change is how the climate of different areas around the globe changes over time, because of the warming of Planet Earth. Such changes are very much evident in the polar oceans and land. 'Climate change' can also occur naturally due to changes in sunlight, the growth of mountains, and the movement of the continents across the earth over time (North Carolina State University).

Scientific Research Findings

The research shows that climate change refers to the change in climates around the world. The increase in global average temperature causes changes in the climate. Climate change means a change in global weather patterns that could affect precipitation averages and extremes. For example, one effect of global warming, one part of the world could be cold, and another part could be so hot that the large polar ice fields melt. In the same area where ice fields were when the ice melts, it leaves a darker open ground. The black ground would absorb sunlight much more quickly than the reflective ice did, leading to excessive heating. Note, heating now in an area where ice once was and freezing.

Another effect of climate change could cause some locations to get more rain, while others will be more likely to have long-term droughts and firestorms. Take, for example, Australia, in one part of the island there is flooding and in the other part, there is no rain but dryness and firestorms with fire tornadoes.

From the subsequent change of global warming, there is climate change, and from the subsequent change of climate, there is a global change. Whatever we call the process, however, our world is getting warmer, and there are variations in the temperature around the world. "Global change can include ecological changes, geological changes, sea-level rise, changes in ocean circulations and acidity, and societal impacts. These changes result in the disruption of our "normal" or expected climate that is likely to occur under global warming and the impacts it would have on life and society" (North Carolina State University).

The Belief Policy About Global Warming

Global warming is what one may call a belief policy or a theory that a large group of scientists holds (Merriam-Webster Collegiate Dictionary 2000) However, since the signing of the agreement to fight global warming/climate change, most scientists and laypeople around the world have accepted the science.

The belief policy or theory of the phenomenon is that humankind has brought about global warming as an observable event on the earth. The theory states that the warming of the planet is a direct effort of humankind. The pollutants from industry, for example, have changed the earth's atmosphere. The pollution and waste that come from the use of the toxic fuels used by humankind have entered the atmosphere. As this toxic waste enters the atmosphere, many chemical reactions take place (Goldenberg 2010).

With the world's dependency, over 80 percent dependent on oil, coal, and natural gas to fuel vehicles, light and heat homes, and drive industry and agriculture, energy is essential for all aspects of human life. Global dependence on these fuels and the concentration of supplies in a few countries mean that energy is a national security issue, and these resources have a connection to geostrategic pressure, instability, and vulnerability, including for the US military (Goldenberg 2010).

Believers in *global warming* suggest that people have taken advantage of our planet by dumping pollution and waste on it. Thus, this observable event—*global warming or climate change*—came into existence. Its effect is the continuous warming of the earth, year by year. Scientists believe this warming of the earth has a direct effect on the weather. It plays an effect on the weather changes in our world, affecting the sky, the land, and the waters of the earth. Therefore, global warming/climate change will continue. Its continuance will yield the occurrences of earthquakes, hurricanes/typhoons, cyclones, volcanic eruptions, tornadoes, tsunamis, floods, fires, snowstorms, and meteorites. Also, this phenomenon causes changes in the ocean and influences animals and humans.

Natural storms will appear unnatural. They will increase in frequency and strength. The intensification of these storms will cause the loss of lives and economic devastation to many countries. The escalation of the heat will cause humans to have physical and psychological changes. Animals too will have drastic changes in lifestyles and the ability to survive.

This scientific teaching holds that if people do not change their current behavior and begin taking care of our planet, it will come to an end. An article from the National Wildlife Federation stated, "Perhaps we should spend less time isolating ourselves from nature and more time trying to reconnect with it, so we may hear, feel, and see what the earth is trying to tell us." *Global warming-climate change* is a topic of interest.

The News Media and the Various Ideas about the Phenomenon of Global Warming – Climate Change

The *Associated Press, USA Today*, the *New York Times*, and The Washington Post have all stated that there is a connection between the extreme heat and the drastic changes in weather around the world that must have to do with global warming. A CNN host referred to the unusual events of global warming as the "apocalypse." The *Huffington Post*, based on their acceptance of global warming/climate change, proposed naming hurricanes and other disasters after climate change "deniers." Stories from the *Associated Press* informed the public that floods, fires, melting ice and burning heat, and the high Arctic, is a 'midsummer breakdown' and "a sign of troubling climate change already underway." Also, the News media, in referring to snowstorms, fires, floods, sinkholes, heatwaves stated that it all signifies, "Weather Chaos, a Case for Global Warming." (Roberts 2010). Many news media professionals and scientists believe that there is a reality about global warming/climate change.

Former President Obama and Vice President Al Gore held strongly in the reality of Global Warming and Climate Change. President Obama held a world conference where world leaders pledged to fight global warming/climate change. Since people throughout the world now know about the science of global climate change, it is a good idea for readers to understand the effects of and the impact on the world. People must know this information so each will be able to make the right choice when the strange and unusual signs are taking place. Vice President Al Gore even wrote a book on Global Warming. However, regarding the US fight with Climate Change, the news media stated that President Trump does not believe in the science of Climate Change. He sees it as a hoax. Yet, many nations whose leaders believe in the 'science's' reality, continue in the fight against the effects of climate change.

Scientists State that Global Warming-Climate Change is Real

Scientists from eighteen scientific associations have published in peer-reviewed journals multiple studies on climate change. All their results point to the consensus that Earth's climate is warming. One outcome says, "Observations throughout the world clarify that climate change is occurring, and rigorous scientific research shows that the greenhouse gasses emitted by human activities are the primary driver." However, let us now investigate the other scientific organizations that state that the "earth is getting hotter." Note their views!

Scientific Organizations and their Views on Global Warming and Climate Change

The findings of the US scientific organizations about the science of global warming and climate change are many. The findings of their investigations are very important in understanding the phenomenon of Global Climate Change. Note some of the findings of the studies.

The American Association for the Advancement of Science found in their 2006 study, "The scientific evidence is clear: global climate change caused by human activities is occurring now, and it is a growing threat to society."

The American Chemical Society, from the 2004 study, states, "Comprehensive scientific assessments of our current and potential future climates show that climate change is real, because of emissions from human activities, and potentially a far-reaching problem."

The American Geophysical Union's studies of 2007, 2012, and 2013 state, "Human-induced climate change requires urgent action. Humankind is a major influence on the global climate change observed over the past 50 years. Rapid societal responses can significantly lessen negative outcomes."

The American Medical Association 2013 study says, "Our AMA ... supports the findings of the Intergovernmental Panel on Climate Change's fourth assessment report and concurs with the scientific consensus that the Earth is undergoing adverse global climate change and that anthropogenic contributions are significant."

The American Meteorological Society 2012 study concludes: "It is evident from the extensive scientific evidence that the dominant cause of the rapid change in the climate of the past half-century is human-induced increases of atmospheric greenhouse gasses, including carbon dioxide (CO2), chlorofluorocarbons, methane, and nitrous oxide."

The American Physical Society 2007 study: "The evidence is incontrovertible: Global warming is occurring. If no mitigating actions are taken, significant disruptions in the Earth's physical and ecological systems, social systems, security, and human health are likely to occur. We must reduce emissions of greenhouse gasses beginning now."

The Geological Society of America revised study 2010 concludes, "The Geological Society of America (GSA) concurs with assessments by the National Academies of Science (2005), the National Research Council (2006), and the Intergovernmental Panel on Climate Change (IPCC, 2007) that global climate has warmed and that human activities (mainly greenhouse-gas emissions) account for most of the warming since the mid-1900s" (2006; revised 2010).

Per the science academies; the international academies: joint statement; and the US National Academy of Science studies agree with the previous findings. These scientific organizations concluded: (**1**) Climate change is real; (**2**) There will always be uncertainty in understanding a system as complex as the world's climate; (**3**) There is now strong evidence that significant global warming is occurring; (**4**) The evidence comes from direct measurements of rising surface air temperatures and subsurface ocean temperatures, increases in average global sea levels, retreating glaciers, and changes to many physical and biological systems; (**5**) It is likely that most of the warming in recent decades can be attributed to human activities (IPCC 2001) (2005, eleven international science academies).

The US National Academy of Sciences study (2005) concluded, "The scientific understanding of climate change is now sufficiently clear to justify taking steps to reduce several greenhouse gasses in the atmosphere."

The US Global Change Research Program 2009 study states that "The global warming of the past 50 years is due primarily to human-induced increases in heat-trapping gasses. Human 'fingerprints' also have been identified in changes in ocean heat content, precipitation, atmospheric moisture, and Arctic sea ice."

The findings of the Intergovernmental Panel on Climate Change stated, "Warming of the climate system is unequivocal, and since the 1950s, many of the observed changes are unprecedented over decades to millennia. The atmosphere and ocean have warmed, the amounts of snow and ice have diminished, therefore, the sea level has risen."

NASA believes, "Human influence on the climate system is clear, and recent anthropogenic emissions of greenhouse gasses are the highest in history. Recent climate changes have had widespread impacts on human and natural systems."

The above findings of the various scientific organizations give proof for the reality of Global Warming and Climate Change.

The US Scientific Organizations Bring Clarity to the Science of Global Warming and Climate Change

The above findings of the US scientific organizations clarify that the warming of the earth is taking place globally. Their findings state the earth is warming. Also, greenhouse gases, because of humankind's pollution, are increasing the temperatures. Therefore, humankind pollutes, the temperature increases globally, and as the temperature increases, the climate changes. This process of climate change has a direct effect on the sky, land, waters, humankind, and animals.

Strange Events are Occurring

Although the earth is warming, science reports that the earth has been colder than the surface of Mars. Also, some parts of countries may experience cold in one section, but in another part, it may have problems with heating. Again, in some countries, one area may experience flooding and in another section of the country, there may be heat, droughts, and fires.

Weather Events in the United States

In the United States, climate changes are presenting in strange and unusual manners. Cold spells broke cold weather records in over fifty cities across the United States. Temperatures have run into the frigid negatives in pockets that typically never fall that low. The authorities of Raleigh, North Carolina stated that the area experienced cold winds, over eighty-six miles per hour. Cold snaps have caused entire rivers to freeze, some of which broke into massive blocks of ice the size of small cars, threatening damage to river assets in several states. Nonstop rain and flooding also have been a problem in some states. Death tolls from cold weather have reached all-time highs across the nation. Also, scientists stated that seismic activity is present near Yellowstone National Park in the United States. This could add to the outburst of volcanic action.

Weather Events in other Areas of the World

Other areas of the world are experiencing extreme unusual weather events. Heavy rains bombarded the Philippines and washed away nearly everything. The storm-torn region caused the displacement of hundreds of thousands of people. Indonesia and even parts of Europe have taken a beating by torrential rains, once again showing a global issue. It was also reported that massive tidal surges were occurring off many coastlines, adding to the problem. January 15, 2014, the temperature in Australia reached 115 degrees, making wildfires a nightmare to extinguish. Many animals died, including one hundred thousand or more bats. In 2014, The Jakarta Post stated how the Mount Sinabung eruption in North Sumatra, which killed sixteen people in Indonesia. After this incident, the authorities raised the status of nineteen other volcanoes.

Also, in other areas of the world, people reported the huge fish and animal die-offs, and the unknown deaths of the whales. Science states that global warming/climate change is causing the strange and unusual events of the weather and deaths of animals (The National Oceanic and Atmospheric Administration, NOAA).

What Science Says About Global Warming and Climate Change Unusual Events

First, since the beginning of the 20th century, scientists have been observing a change in the climate. The findings state that there are some natural causes of climate change, but humans are also responsible. Humans have contributed to the increase of heat-absorbing greenhouse gasses in our atmosphere since the industrial revolution. The ever-increasing amount of these gasses has directly led to more heat kept in the atmosphere and thus to increasing global average surface temperatures (World Meteorological Organization).

Second, the findings of the various research projects yielded information that clarifies that temperature change is the main component and is responsible for many of the strange and unusual events. These events will continue to take place in the sky, on land, in the waters, and with the weather. Also, "over the next several millennia, global warming could be irreversible. Even if people drastically reduce emissions, global temperatures will remain close to their highest level for at least one thousand years" (Solomon, et al. 2008).

Third, the studies of the scientific community on the effects of such drastic warming of the earth say the changes in the temperature will have effects on agricultural and forestry management at Northern Hemisphere higher latitudes. Therefore, spring planting of crops should take place earlier, and there will be alterations in the disturbances of the forests because of fires and pests. Some aspects of human health, such as heart-related mortality, will rise in Europe. Changes in infectious diseases will increase in some areas, and the Northern Hemisphere high and mid-latitudes will experience an allergic outbreak from the pollen. Such conditions will affect human activities, such as hunting and travel over the snow and ice in the Arctic regions of the world (IPCC 2007).

The United States' Intervention in Global Warming-Climate Change Prevention Effort

Despite all the controversy and hype that climate change has generated, there now exists an overwhelming body of scientific evidence that shows the problem of climate change is real and that its effects are already on a global scale. It also involves economic, sociological, political, psychological, and personal issues, making this a topic that affects every person on earth, now and in the future (Gines 2011).

As noted in the United States scientific organizations' study outcomes, climate change extends far beyond the physical sciences to affect lifestyles, cultural values, political systems, economics, and health. Aspects such as international cooperation, journalistic balance, human psychology, international policy, national security, socioeconomic impacts, agricultural conservation, health care, and economics of mitigation, climate modeling, and error amplification are real. There is no time for "media sensationalism." It is important to "enlighten and empower readers to look at the decisions that people must make to mitigate climate change problems before irreversible damage" (Gines 2011).

Former President Obama's Contribution to Reality of Global Warming/Climate Change

President Obama of the United States saw climate change as a serious issue. Under his leadership, he laid down preparations for the United States to deal with global warming/climate change. His belief was unyielding, and he took the lead among the world leaders to do something about it. President Obama's belief was, "Climate change is real and dangerous." So, he took bold action to reduce carbon emissions and help the environment.

According to Graves (2013) President Obama's strategy for fighting global warming/climate change during his second term, rested on three pillars: (1) cutting carbon pollution in America, (2) preparing the United States for the impacts of climate change, and (3) leading international efforts to cut global emissions (Graves 2013).

The News Media and the President's Strategy for the Climate Change Fight

On June 25, 2013, President Obama invoked his executive authority to undertake a slew of measures aimed at curbing climate change and preparing America for its harmful impacts. President Obama echoed his feelings and beliefs in a speech at Georgetown University. Note, "The question is not whether we need to act. The question is whether we will have the courage to act before it's too late" (*Huffington Post*).

In President Obama's second inaugural address, January 2013, he challenged the people of the United States to (1) recognize their obligations in the fight, (2) respond to the threat of Climate Change on behalf of their children, (3) the denial of the science of climate change does not mean it is not real, (4) consider what was happening to America with the 'raging fires and crippling drought and more powerful storms,' (5) action by the American people will "preserve our planet, commanded to our care by God." President Obama's initiatives about 'clean-energy' initiatives to fight against climate change received stiff opposition by the United States Republican Party (Graves (2013).

Vice President Al Gore's Contribution to Reality of Global Warming/ Climate Change

Former vice president Al Gore of the United States contributed much to the global climate change fight. He led many campaigns, gave many speeches, and wrote a book on global warming. In his book, "Our

Choice: A Plan to Solve the Climate Crisis", Mr. Gore stated the serious damage that is already done to the global environment and to healthy climate balance. He felt that the damage has a paralyzing effect on the earth and its people. Therefore, he laid out solutions for people to overcome present obstacles. He felt that people should, (1) change their thinking, (2) understand the real cost of carbon and the political constraints, and (3), understand the meaning of the Bible message of Ecclesiastes 4:9–10, "Two are better than one because they have a good return for their labor: If either of them falls, one can help the other up." Meaning if Americans work together, they will be able to overcome the dangers of climate change (Gore, 2009).

The work by the former president and Vice President in America appears to have caused other leaders to join in the global climate change fight. Years after President Obama's initiatives on climate change, the UN Secretary-General voiced the reality of global climate change, says Time Magazine.

The UN Secretary-General's Warning About the Reality of Climate Change

According to Time Magazine's interview with the United Nations Secretary-General Antonio Guterres, he stated that Global Climate Change is Heating the Planet Earth. Also, he said, "regardless of the lack of interest by some of the world leaders of Global Climate Change, there is a dangerous hot climate and political tensions." He further noted that "there are (1) rising seas, (2) fleeing residents, (3) disappearing villages, and (4) our sinking planet." Also, he warned that "These extreme weather events are just the tip of the iceberg. And that iceberg is also rapidly melting…bold action and much greater ambition." The Secretary feels that there is a need for a new energy source, the planting of millions of trees to reverse deforestation and remove carbon dioxide from the environment. The Secretary-General shows 'the Sinking Planet Earth,' as he stands in knee-deep water (Time Magazine, 2019).

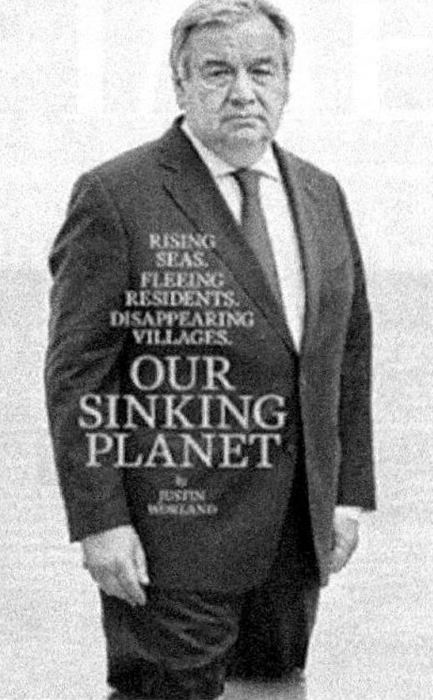

Regarding all that is going on with and among the proponents who are engaged in the fight against global climate change, there are those who oppose science. The earth will continue to warm, and storms will continue to increase in their frequencies and strengths. For example, disbelief in the science of Global Climate Change will not stop the fires of the world nor the increase of the CO2 levels. It will not stop the mass deaths of animals that fly in the sky, roam the land, or swim and live in the sea. The disbelief will not stop the natural occurrences of earthquakes, hurricanes/typhoons, cyclones, volcanic eruptions, tornadoes, tsunamis, floods, snowstorms, and meteorites. Most importantly, the scientists stated that the escalation of the heat will cause humankind to have many health problems.

Science's predictions about Global Warming/Climate Change and its drastic effects on Planet Earth is a must to read about. This information affects you! So, the following chapter explains it all. READERS MUST KNOW THE TRUTH. Let's Read!

Chapter 7

Science Predicts How Global Warming/Climate Change Will Affect Planet Earth

Photo shows 'The Puzzle'

Regional effects of global warming/climate change vary in nature. The world's average temperature is rising. When global temperature changes, there will be changes in climate across the earth. However, people will notice that the changes will not have uniformity. They will see the effects on land more than in the oceans. They will also see the changes more quickly in the northern high latitudes than in the tropics.

Impacts on Africa may be different from those in Asia, Australia and New Zealand, Europe, Latin America, North America, the polar regions, and on small islands. But global climate changes will have a great effect on birds, land animals, and aquatic animals. The effects also will have a great impact on humankind and its environment (Wikipedia, 2019).

In presenting the Scientific Predictions of Global Warming/Climate Change, this chapter presents and discusses eight main areas. These are: (1) Global Warming/Climate Change Effects on the Regions of the World and the Outcomes; (2) The Effects on Earth's Biological Balance between Plants, Animals, and Humans; (3) The Effects on the Health and Social Systems; (4) The World Health Organization Predicts Seven Important Areas that Global Warming's Climate Change Will Affect; (5) Science Predicts that Global Warming's Climate Change Will Influence the Military Might of the World: (6) Science Predicts that Global Warming and Climate Change Will Influence the Seas and Oceans of the World; (7) Global Warming/Climate Change Effects on the Weather; (8) Concluding Thoughts.

Global Warming/Climate Change Effects on the Regions of the World and the Outcomes

Impact on Africa

Science predicts that because of poverty, political conflicts, degradation of the ecosystem, and low adaptive capacity, Africa is one of the most vulnerable continents to climate variability. So, science projected that by 2050, between 350 million and 600 million people will experience the stress of three main factors because of climate change. These are: (1) a severely compromised agricultural production – shortage to food; (2) sea-level rise will affect low-lying coastal areas; (3) health problems because of a shortage of food, water, and a heightened impact of diseases. These conditions will put the inhabitants into a survival mode, which will create wars.

Impact on Asia

The melting of the glaciers in Asia will cause flooding especially in the delta regions in the south, east, and southeast. The flooding will affect drinking water. Crops could decrease by 30 percent. Because of the huge population, the reduction in food and water will cause drought and famine.

Impact on Australia and New Zealand

Similar problems will also affect the countries of Australia and New Zealand. By 2020 there will be a significant loss of fish and other creatures that provide wealth, health, and food. Severe storms and coastal flooding will affect the infrastructure. Also, severe heat waves will cause droughts and fires which will cause agricultural failure. Science states that by 2030, water and food shortage will be a factor both in Australia and New Zealand.

Impact on Europe

Europe will undergo severe heat waves which will cause the temperature to rise. The effect will reduce the agriculture, energy, tourism industry, and economics. Also, there will be winter floods, landslides, and earthquakes. These factors will produce endangered ecosystems, science predicts.

Impact on Latin America

First, the rising temperatures will cause the glaciers to melt. This event will influence energy production and flooding. The flooding will cause the rise of sea levels, the displacement of people from the lower coastal areas to higher grounds. *Second*, the rising temperatures will affect the soil moisture, dry up the water tables and the saline will worsen the grounds. Therefore, the production of livestock, maize, and coffee, science predicts a decrease. These events will cause an economic failure, food and water shortages in Latin America.

Impact on North America

The scientific community is predicting that temperatures in the western mountains of North America will decrease in snowpack areas, and therefore, there will be an increase in winter flooding. There will be a reduction in the summer flow, which will cause a water problem. There will also be disturbances from pests; diseases will increase, and the risk of fire will increase. Crops will face significant challenges as heat waves increase. The older population will be at risk of increased heart problems. Also, climate change will be a mechanism for increasing stress within the coastal communities and living areas. This stress among individuals will come from pollution and all the other problems with water, heat, and shortages with food.

Impact on Small Islands

Changes in the sea level will affect small islands. With high tides, there will be a steady deterioration of their coastal regions. Also, destruction will come to the beaches, where tourists come to relax. With the destruction of the coastal areas, the island communities and economic structures will fail. People will flee, as the increase of hurricanes, cyclones, and such will destroy these low-lying islands. People, in the long run, must escape, which will cause a massive displacement and abandonment of the islands (Tsosie 2007).

The Effects on Earth's Biological Balance between Plants, Animals, and Humans

There is a scientific balance between humans, animals, and plant life. People and animals depend upon the trees for oxygen, which they give off. The trees depend upon humans and animals for carbon dioxide, which they breathe out. Global warming/climate change will interrupt this balance.

Science Predicts that the Balance Between Plants, Animals, and Humans is Changing

First, Global Warming/Climate Change is already affecting the balance between plants, animals, and humans. The *National Geographic* (2009), states that the industrial areas of the world affect the sky with pollution. The pollution "magnetizes" the leaves of the trees and destroys them. The destruction of the trees will affect the balance of transfer between oxygen and carbon dioxide.

Second, the leaves of the trees give oxygen essential for sustaining life, both human and animal, on earth. The carbon dioxide which people breathe out supports life for the trees in the land (Lovett 2009). Findings say that there will be a scarcity of oxygen and problems with carbon dioxide, ultimately leading to the deaths of humans, animals, and plants.

Science Predicts that Climate Change is having a Tremendous Effect on Species and the Extinction of Animals

Changes in the melting glaciers, the shrinking rainforests, and the rising seas, are giving a proper diagnosis of the planet, but there seems to be no cure for the sickness of global warming/climate change (Miller 2009). Global warming/climate change will spread. Such predictions of science, which many people feel have already begun, do not give a good outcome for humankind, and all other living creatures.

First, regarding animals, "Climate change also has a tremendous effect on species and extinction of animals (Sahney, Benton, & Falcon-Lang 2010). The heat will have a direct effect on the animals' habitat and will interfere with their survival. Also, the heating of the earth will affect the animals' protection from natural predators and their food supply from competitors (Yohe and Parmesan 2003). The temperature of the soil causes the animals to come out of their living quarters on the ground. They become exposed to predators. Their methods of hunting for their survival are about to become threatened, and they eventually will lose their lives. After many years of trying to survive this way, they will become extinct.

Second, there are also predictions about the heating of the earth and the melting of areas where ice is always present. The Antarctic Peninsula is experiencing one of the fastest rates of regional climate change on the earth. Thus, there is a collapse of the ice shelves. Such collapsing has a *dire effect on glaciers and animals that live in these regions.* This threat is because winter sea ice in the surrounding oceans is decreasing by 10 percent per decade. Also, the sea ice and permafrost will have negative impacts and could damage the infrastructure of these areas.

A recent article by the Associated Press/ NPR, dated March 25, 2022, stated that,

> A previously stable ice shelf, the size of New York City, collapses in Antarctica. The satellite image shows the main piece of c-37 close to Bowman Island. Scientists are concerned because ice shelf the size of New York City collapsed in East Antarctica, an area that had long been thought to be stable…It happened at the beginning of a freakish warm spell last week when temperatures soared more than 70 degrees warmer than normal in some spots of East Antarctica (Associated Press -NPR, 2022).

Scientists are troubled over the fact that the process "would raise seas across the globe more than 160 feet (50 meters)." Also, it must be noted that with this melting of the ice glaciers, there will be much flooding on the land and erosion of shores, and damages to the landmass and homes etc.

Third, the melting of the ice in the polar regions will add to the rising sea levels, which will contribute to flooding. Because of the warming of the oceans and the melting of the ice, marine ecosystems and habitats will be at risk. Once the heated oceans break down the protected climatic barriers, arctic creatures will cease to be. Science shows that the warming of the oceans will have a biological effect on the animals that live there (Rosenzweig et al. 2007).

The literature suggests that because of global warming/climate change, the polar regions of the earth are now becoming a target for global warming. The animals of these areas that depend on earth's colder weather will die and become extinct. The *National Geographic* article, "The Arctic Largely Ice-Free in Summer within Ten Years?" predicted the melting of the ice from the polar regions within a decade (2009). Another article by Carroll (2010) predicted the same outcomes. The conclusion means that soon there will be no home for animals such as the polar bear and others who depend on the polar regions for survival. They will all die out.

See the photo below showing the melting ice in the polar regions of the earth.

Photo showing the results of the melting of Ice as predicted by Science

Fourth, various studies have agreed on global warming/climate change affecting the biological changes of the ecosystems, plants, animals, and humans. However, they believe that as time goes on, climate change will affect the present ecosystems. This change will alter the structure, reduce biodiversity, and perturb the functioning of most ecosystems, and compromise the services they provide (Chika, 2011). Losing so many systems of this world will eventually cause substantial damage to or complete loss of some unique ecosystems, and probable extinction of some critically endangered species of the planet's animals.

The Effects of Global Warming and Climate Change on the Health and Social Systems

According to scientific studies, they report that Climate Change will have a great effect on Human Health and Human Social Systems. The findings of study one on people's health because of global warming/climate change revealed that people who live in regions of the world with perilous human security, and pervasive poverty, geographic disadvantages will suffer. Also, people who live in areas where factories and other contributing factors are present that produce greenhouse gas emissions will also suffer from health issues (Gardens for Health International 2015).

A *second* study revealed that people will experience, (1) heightened emotions from the effects of the heat, (2) the shortage of food, (3) drought, (4) the migration effect in seeking food and water will increase stress, (5) climate related-conflicts and the adjustments from travel will add to the psychological and sociological problems of people (Doherty & Clayton 2011).

The *third* set of studies by the World Health Organization (WHO) sums up the findings how the increased temperatures will affect human physical and emotional wellness as it relates to, (1) water supply, (2) food supply, (3) energy, (4) migration and conflict, (5) economics, and (6), health. The description of each of these areas will inform how global warming/climate change will affect people as the earth continues to heat up. During this discussion, sometimes, the chapter may refer to Global Warming/Climate Change as GWCC.

Global Warming/Climate Change Effects on Water Supply

"Push aside bread; water is the true 'staff of life.' Water carries oxygen to our body's cells, makes up more than half of our body weight and we can't live for more than a few days without it" (Nath 2012). This means, without water, people will die. Read a few facts about water as it refers to human health:

- 66 percent of the human body is water.
- 75 percent of the human brain is water.
- By the time you feel thirsty, your body has already lost over one percent (1%) of its total water.
- A person can live for over a month without food, but only about three days without water.
- Globally, 69 percent of withdrawn water is for agriculture, 23 percent is for industrial, and 8 percent is for domestic (befoundalive).

The Intergovernmental Panel on Climate Change (IPCC) report on water states:

1. The freshwater resources that humans rely on are highly sensitive to variations in weather and climate."
2. As the weather warms, it changes the nature of global rainfall, evaporation, snow, streamflow and other factors that affect water supply and quality.
3. Rivers and streams, from which humans get their water supplies, must depend on the water from the melting snow. Therefore, warmer temperatures will interfere with the process of rainfall and snowfall. There will be an increase in winter flooding and a reduction in flowing water in the summer. This process will cause the seas to rise and saltwater to enter the underground water table and fresh streams. Such a process will decrease the amount of freshwater available for drinking and farming.
4. The warmer the water becomes, the more chances of poor quality and polluted water. These conditions will add to the poor health of people. Death will be the final stage of short supplies of water.
5. Lack of water means the inability to irrigate crops in countries of the world. Therefore, food supplies will decrease and famines around the world will increase. These shortages will affect People's health.
6. Increased heat will increase evaporation which will reduce the effectiveness of reservoirs which means less water for human consumption. The human demand for water will grow since the population is growing. If water is not available for cooling and hydration, death will be imminent.
7. Lack of water means poor health practices. Water helps to keep people clean, washing away dirt and filth that could lead to infectious diseases. Therefore, water shortage will affect people's health.
8. As the need for food, the need for water will also be a determining factor for starting a war.

An article on Water, International Peace, and Security, states that "water scarcity, sped up by climate change, affects water availability and may threaten peace and security" (Tignino 2010).

Global Warming/Climate Change Effect on Food Supply

The scientific community's picture of global warming/climate change and its effects paints a disturbing picture for this planet, with chances for water shortages, water pollution, and shortage of food supplies. Together, these shortages will become a global conflict. Also, such shortages will put the world into a panic mode.

In an article, "Development: Resource Crunch Signals Larger Ecological Crisis," Caldwell (2010) expressed the seriousness of the conditions of not having enough food and water. *First*, the article stressed that by 2030, almost four billion people will live in areas of high-water stress by 2030. *Second,* with agriculture accounting for 70 percent of freshwater withdrawals, there are also clear linkages between water scarcity and food production. *Third,* the World Economic Forum (WEF) reports that food demand will grow 70 to 90 percent by 2050.

Science is predicting that GWCC will one day cause the earth to be short of water and food. Therefore, nations are looking for some secure means of surviving, such as an impending famine (Easterling et al., 2007). By the year 2050, science predicts that food and water deficits will cause much confusion. Governments of the world will have to feed over four billion people. India and China will have the greatest demands on the world for supplies because they alone will need half of the world's supplies for themselves. With such requirements, this could be a means of increasing conflict among the nations. Many people feel that if we do not face this problem now, it may lead to another world war.

Global Warming/Climate Change Effect on Energy

The International Energy Agency's World Energy Outlook 2008 notes that world energy demand will grow 45 percent by 2030. India and

China alone will be responsible for over half of the increase in energy demand by 2050. It is unclear how the providers will meet the demand for energy for the expanding populations of the world.

Water, heat, and wind, each produces energy. Lack of energy, especially electricity, means that lighting, cooling, and various means of cooking and relaxation, will not be available. The availability of such sources will be only for those who can afford it. When the water supply is becoming low, systems that depend on hydropower energy will not function. Therefore, a water shortage will mean a shortage of electrical energy. With a rise in the world's population, the problem will increase. Also, electricity comes from solar panels and wind power. The unpredictable behavior of the weather because of global warming/climate change could cause instability with the wind. Lack of wind means the wind turbines will produce no energy.

The prediction results are clear: there will be great stress on the people of the world when the world becomes overpopulated and food, water, and energy are in short supply. People also need other sources of energy. A crowded and heated world means that people will need air conditioning in homes. Poor people will be vulnerable, and such vulnerability leads to poor health. People will die from heat stroke or heart conditions. Such situations no doubt would cause nations to fight for survival. Demands can foster conflicts, and conflicts can lead to war, and war kills. We see the coming problem and the effect on the world and its people that science is predicting global warming and its climate change will cause.

Global Warming/Climate Change Effect on Migration and Conflict

Note this prediction: "future climate change will bring wetter coasts, drier mid-continent areas, and further sea-level rise. Therefore, "environmental degradation, loss of access to resources (e.g., food, water resources), and resulting human migration could become a source of political and even military conflict" (Desanker, et al. 2001). The process of migration involves many social, economic, and political policies. So, think of a time when there is a scarcity of food, water, energy, poor health, and all the other problems from climate change. Accepting immigrants who are fleeing for their lives may become a problem with a country's people and government. This is a problem that global warming/climate change poses in regard to migration.

A 2014 study by Ranson, associated higher temperatures with a greater likelihood of violent crime and predicted that global warming would cause millions of these offenses in the United States alone during the twenty-first century. Note these statistics emerged from the analysis based on a 30- year panel of monthly crime and weather data for 2997 US counties. The monthly crime-based statistics on weather results show that there is a strong positive effect on criminal behavior, with little evidence of lagged impacts. Between 2010 and 2099, climate change will cause an additional 22,000 murders, 180,000 cases of rape, 1.2 million aggravated assaults, 2.3 million simple assaults, 260,000 robberies, 1.3 million burglaries, 2.2 million cases of larceny, and 580,000 cases of vehicle theft in the United States. If this is only in the United States, think about the other countries of the world and the effect of global warming/climate change will have on people.

Global Warming/Climate Change Effect on Economics

First, the scientific community is predicting that socio-economic factors will contribute to global losses. The reason—population growth, changes in precipitation, and flooding events. changes over land areas from one type of vegetation to another would influence the production of goods and the gross domestic product. History has shown that the global losses because of extreme weather-related events since the 1970s have caused rapidly rising costs. But there are more problems to come. For example, the Wikipedia research on this subject stated, "climate change in the twenty-first century is likely to adversely affect hundreds of millions of people through increased coastal flooding, reductions in water supplies, increased malnutrition, and increased health impacts." These are a few of the climate change problems which will affect the world economy.

Second, hurricanes, wildfires, and drought pose enormous risks to insurers. Some media stories stated how several national firms have stopped writing new policies altogether or pulled back in other ways. For example, insurers are afraid of disasters like Hurricane Andrew in 1992, which cost the industry an estimated $25 billion. Since 1992, many other hurricanes have caused great economic damage to countries. Like Hurricane Darian which almost wiped out two of the Bahama islands in 2019, caused great economic damage to the Bahamas.

Third, the increased heating of the earth and the unbearable temperatures are also causing economic burdens. Some governments have to spend much of their resources on protecting their people. For example, an exceptional heatwave in Europe in 2003 took at least thirty-five thousand lives. Afterward, French cities set up air-conditioned shelters and identified older people who would need transportation to the shelters. Similarly, after a tropical storm killed 500,000 people in Bangladesh in 1970, the government developed an early warning system and built basic concrete shelters for evacuated families. These are just a few of the economic burdens that global warming/climate change is putting on countries of the world.

Global Warming/Climate Change Effect on Human Health

First, the shortages of food and water supplies, and the interruption of energy will compromise human health. Certain groups of people are sensitive to climate change impacts, such as the elderly, the infirm, children and pregnant women, native and tribal groups, and low-income populations will have health problems and therefore, death (The United States Environmental Protection Agency). Besides the above causes for health problems, the increased temperatures alone will become unbearable for humans and will cause people to die (Dunne, et al. 2013).

Second, the World Health Organization (2009), states that global warming/climate change was responsible for diseases such as diarrhea, malaria, and dengue fever deaths worldwide in 2004. Projections continue to show that there will be adverse health effects from rising temperatures, especially in developing countries. Also, 160,000 deaths since 1950 are direct results of climate change. The current and future impact of global warming/climate change will continue as a human emotional threat. Such a threat will affect health by producing psychosocial problems, especially among the old and the poor.

Third, military planners see global warming as a threat multiplier. Therefore, poverty, food and water scarcity, diseases, economic instability, and the risk of natural disasters caused by climatic conditions may threaten the stability in much of the world" (2013 Spaner & LeBali).

Global Warming/Climate Change and its Influence on the Military Might of the World

The military might of the world in question are those of the United States, Canada, the United Kingdom, Russian, and China. This section of the chapter presents and explains each of these major country's military actions toward Global Warming/Climate Change.

Global Warming/Climate Change Effect on the United States' Pentagon and the Military

Since 2010, the United States has recognized that Global Warming and Climate Change posed a serious problem to the nation. Suzanne Goldenberg (2010) stated that (1) the Pentagon ranks global warming as a destabilizing force, (2) it will put US troops at risk around the world, (3) it will direct military planners to keep track of the latest climate science, and (4), it allows the military to factor global warming into their long-term strategic planning.

The Pentagon has changed over thirty US bases because rising sea levels, severe heat waves, and fires have threatened the locations. The CIA and the environmental experts in Washington are working together to collect intelligence on climate change to gain satellite images of glaciers and Arctic sea ice, and they are seeking methodologies to reduce greenhouse gas emissions. Besides the science of and preparation for global warming/climate change, US troops are feeling the effects of climate change directly. There is an increased demand for the US military to respond to humanitarian crises or natural disasters.

The United States Department of Defense (DoD) Mandate for Global Climate Change

Since the above precautions, the U.S. Department of Defense (DoD) released a congressionally-mandated report detailing the challenges climate change poses to the U.S. military.

First, the report cited increased exposure to recurrent flooding, drought, desertification (a process by which fertile land becomes desert), wildfires, and thawing permafrost. The report highlights how climate change affects U.S. military readiness to respond to national security emergencies.

Second, the report included a list of selected climate change events that compromised mission-related activities at military installations because of environmental vulnerabilities. Therefore, the military adopted policies to mitigate future damages.

Third, in showing how global climate change has threatened the military, the report presents 79 priority American domestic installations and their critical operational roles.

Fourth, the report concludes that Global Climate Change is an important and tangible threat to the U.S. military.

Fifth, the report noted that hurricanes and wildfires were also on the critical list.

Sixth, the melting of ice interferes with protection and yes, the vulnerability of different countries having excess to an enemy's country or vice versa. Also, the delivery of aid will also be a problem in emergencies (cfc.org).

The above are areas in which the US Military has acted pertainingto the ongoing threat of global warming/climate change.

Global Warming and Climate Change Effect on Canada's Military

First, Global Warming/Climate Change has put the Canadian Armed Forces on a continuous and careful watch. Defense Minister Rob Nicholson stated that "the Canadian Armed Forces are watching the changing climate in the North very, very carefully." There is an impact of climate change on naval routes through the Arctic which has affected the military. Therefore, Canada, realizing the vulnerability and risks and the impacts of climate change on the Arctic, has focused on specific security attention for the area (Gaines 2011).

Second, Global Warming/Climate Change is pushing Canada's Armed Forces to the limit. Members are constantly responding to the increasing number of climate-related events such as floods and fires. There is a feeling within the military that these incidents are increasing each year. Therefore, the military brass understands that Global Warming/Climate Change has placed a demand for additional men and women to handle the climate change-related crises. The demand has also increased the need for additional training of the soldiers to deal with fires and floods. Therefore, Global Climate Change already has placed on the Canadian Military forces the need for engagement in constant military preparedness.

Third, Global Warming/Climate Change has forced the formation of military alliances between Canada's and the United States military forces. The alliance will consider resilience building, humanitarian crisis response, deployments, procurement, training, equipment, and energy consumption. The Department of Defense of the Canadian Forces feels that it is imperative to integrate climate change into all policy and planning frameworks (Gaines 2011).

Fourth, the Canadian federal agencies, levels of government, and Aboriginal groups are all geared to "protect the environment." They carry out this type of protection by the cleaning up of chemical spills and other sorts of disasters with which they get involved with the local authorities. Besides all these local activities, the naval and military forces also carry out local activities. Besides, the authorities are holding meetings and educating the people to bring about awareness of climate change (Boutilier 2014).

Global Warming/Climate Change Effect on the United Kingdom's Military

First, the UK's foreign secretary described Global Warming-

Climate Change as "perhaps the 21st century's biggest foreign risk for global war, a risk we should think seriously about preventing right now." Therefore, the UK is preparing its armed forces for any violent conflict posed by Global Climate Warming/Climate Change.

Second, Global Warming/Climate Change has caused the UK military to be ready to encounter activities of migration, environmental degradation, and resource scarcity that may cause conflict (Governance and Social Development Resource Center, 2007). If such future conflicts may arise, the British military has a heavy responsibility, not only to protect and defend its homeland but also its waters, food supplies, other livable resources, and its Overseas Territories, in case of storms which Global Climate Change brings.

Third, Werrell & Femia's 2015 article's findings made it clear that the UK's National Security Strategy & Strategic Defense & Security Review must deal with Climate Change. So, in the UK's strategic document, the plan gives full support for the climate change fight. The UK's military carries out greater involvement in the actual responsibilities of the Global Climate Change events. The officials see climate change as a grave threat to the United Kingdom's security and economic resilience equal to terrorism and cyberattacks. To them, climate change is one of the greatest risks because it poses a global threat. Therefore, the UK involves the three branches of the armed forces—the navy, army, and air force—in the climate change fight.

Global Warming/Climate Change Effect on the Russian Military

The 2012 Bulletin shows that the consequences of climate change and its posture in the Arctic has heavily influenced Russian military planning. *First,* Global Warming/Climate Change has forced the Russians to engage in operational regimes in the Arctic Zone. *Second,* it caused the Russian military to engage in incentives to reduce energy consumption but there is no concrete action regarding the reduction of greenhouse gas emissions. *Third*, the science of Climate Change has not caused the Russian military to prepare for the fight against climate change as it relates to any of their military bases. Nor has it stirred them about the precautions to combat the rising sea levels. Unlike many other nations' efforts with their militaries, the Russian national security strategy in the face of a warming Arctic continues to,

1. Use the Arctic Zone of the Russian Federation as a strategic resource base.
2. Safeguard the Arctic as a zone of peace and cooperation.
3. Use the Northern Sea Route as a national integrated transport communication system.
4. Create a unified regional system of search and rescue. Also, to have the preemption of technogenic catastrophes. In addition, be able to eliminate the consequences of catastrophes and have a good rescue service.
5. Accept a major role in domestic disasters possibly related to climate change with regards to war, and disaster in Russia (Brzoska, 2012).

Apparently, Russia may not have fully accepted the science of global climate change. Therefore, they are not making much preparation except for being able to respond to a war caused by Global Warming/Climate Change.

Global Warming/Climate Change Effect on the Chinese Military

China, over the last few years, has mentioned climate change as being among a security issue. But China in order of importance identifies its country's threats as, terrorism, economic insecurity, climate change, nuclear proliferation, insecurity of information, natural disasters, public health concerns, and transnational crime. Global warming/climate change is No. 3. Although the science of climate change is not the No.1 important factor, yet China's interest is in building its military. They are spending more than the United States on such military buildup. Other studies show that China feels accepting climate change as a security issue might be a justification for Western military interventions in future crisis situations (Freeman, 2010).

Another study says China seems to take a 'wait-and-see' approach rather than openly positioning itself to GWCC science (Jakobsen, 2010). However, China continues to prepare; becoming militarily ready for whatever global climate change may present. China has rejected the science of Global Warming/Climate Change and therefore, does not see it as a priority. However, the country continues to build its military.

Science Predicts that Global Warming and Climate Change Will Influence the Seas and Oceans of the World

Global Warming and Climate Change Effect on the Temperatures of the Oceans

From the 2013 Mud Report, note some of the findings that relate to the oceans.

First, during the past 25 years, satellites have measured a 4 percent average rise in water vapor in the air column. The more water vapor, the greater the potential for intense rainfalls.

Second, by the end of the century, the average world temperature could rise anywhere from three to eight degrees Fahrenheit—depending in part on how much carbon we emit between now and then.

Third, one of the biggest wild cards in our weather future is the Arctic Ocean, which has lost 40 percent of its summer sea ice since the 1980s. Autumn temperatures over what is now open the Ocean have risen 3.6 to 9°F, as the dark water absorbs sunlight that the ice once bounced back into space.

Besides the Mud Report (2013), scientists say that the role of the oceans in global warming/climate change is very complex. Note:

(1) The ocean serves as a sink for carbon dioxide.
(2) It takes up much of this gas that will remain in the atmosphere.
(3) It also increases the CO2 levels, which lead to ocean acidification.

(4) If the temperature of the ocean increases, absorbing CO2 becomes involved. Then, Global warming/climate change is projected to have an ongoing effect on the sea by raising the sea levels because of the increase of heat, melt much of the ice, and cause some large-scale changes in ocean circulation.

To support the above scientific findings, research by Fischlin et al. (2007) projected that, (1) future ocean acidification, (2) changes of CO2 and (3) heating of the oceans would influence marine organisms and shellfish. The southern oceans will feel this effect. The marine animals will have a problem sustaining life.

Global Warming/Climate Change Affects the Oxygen Levels in the Oceans

A few studies done on oxygen depletion as it relates to global warming/climate change have resulted that show a depletion of oxygen in the oceans. The depletion of oxygen will affect and cause adverse consequences for ocean life. Crowley & North (1988) study stated:

> Terrestrially induced climate instability is a viable mechanism for causing rapid environmental change and biotic turnover in earth history ...living organisms—those in the sea, on land, and in the sky— will have problems with maintaining their biotic environment as it relates to food resources and symbiotic relationships.

This will mean death to many of the creatures, as we already have seen in the reports of birds, fish, and other creatures that are dying without explanation. Mentioned most times was the lack of oxygen. Shaffer, et al., wrote an article on "Long-term oxygen depletion in response to carbon dioxide emissions from fossil fuels." The findings state (1) an adverse effect on marine life – increased mortality events, (2) long-term ocean oxygen depletion, and (3), a significant expansion of ocean oxygen-minimum zones for scenarios with high emissions or high climate sensitivity.

Global Warming/Climate Change Will Melt the Ice in the Polar Regions, Causing Massive Flooding

According to Bindoff & Willebrand (2007), there is strong evidence that global sea-level rose gradually during the twentieth century. This study concluded that between the mid-nineteenth and mid-twentieth centuries, the rate of sea level rise increased. The two main factors that contributed to this increase are thermal expansion (as ocean water warms, it expands), and the melting of the ice because of the warm seawater.

The melting of the ice glaciers means a rise in sea levels. Increased sea levels mean a greater possibility of flooding. Because the water is already warm, there is a bio-effect on the animals of the oceans. Death to these creatures will be imminent.

Global Warming/Climate Change Effect on the Weather

Unusual Heat Production is Creating a Great Effect in Various Parts of the World

First, the National Geographic (2012) noted that in theory, extra water vapor in the atmosphere should pump heat into big storms such as hurricanes and typhoons, adding buoyancy that causes them to grow with power. Some models have predicted that global warming could increase the average strength of hurricanes and typhoons by 2 to 11 percent by 2100.

Second, a hotter and wetter atmosphere promotes more severe thunderstorms that are producing many tornadoes. Thus, the US is reporting more tornado activity. Spring of 2011 was one of the worst tornado seasons in US history, with monster twisters roaring through Tuscaloosa, Alabama, and Joplin, Missouri. Per Gerald Meehl, a senior scientist at the National Center for Atmospheric Research in Boulder, Colorado, "global warming had changed the odds for extreme weather."

Third, National Geographic stated that global warming causes the addition of more carbon dioxide to the atmosphere. The carbon addition makes the atmosphere a little bit warmer and shifts the odds toward these more extreme climate 'changes' with strange and unusual events with rain, temperatures, and storms.

Fourth, scientists are now viewing variations in the weather by the amounts, frequency, and type of precipitation. Projections of future changes in precipitation show an overall increase in the global average. Dry areas at present will become even drier, while regions that are wet will become even more humid (Intergovernmental Panel on Climate Change 2007).

Fifth, scientists also project that variations in the frequency and intensity of some extreme weather events will happen by 2035. Therefore, during the years 2081 to 2100, the intensity of temperatures will cause changes in storms, an increase in the drought, and reductions in soil moisture (Stocker, et al. 2013).

Sixth, scientists project that the Arctic summers could be ice-free as early as 2025– 2030. This event will cause the animals that live in the Arctic regions to be in jeopardy and probably lose their lives (Met Office 2012).

Seventh, since 2014, the projected effects of global warming and climate change the world will experience (The Guardian's John Vidal (2014). Many areas of the world have already experienced the projected effects and there are more to come. For example, (1) there were heat waves in Slovenia and Australia, (2) snow in Vietnam, (3) the return of the polar vortex to North America, (4) Britain had its wettest winter in 250 years, (5) the temperature in parts of Russia and the Arctic was fifty degrees above normal, (6), the Southern Hemisphere saw the warmest start to the year ever recorded, and (7), the high temperatures affected millions of people in cities in Brazil and South Africa.

However, the effects of the heat continued its effect on the world. For example, (1) within the United States, the freezing of Lake Michigan, (2) North America experienced a seemingly endless winter of snow and ice, (3) both equatorial and polar regions experienced extremes, (4) there was unusually heavy snowfall in the Southern Alps, (5) the monthly temperatures were high from eastern Mongolia to eastern China, (6) in the Southern Hemisphere, Australia, Argentina, and Brazil experienced extended heat waves, and (7), the unusually cold weather in the eastern United States coincided with severe storms in Europe (Vidal

2014).

Eighth, the UN's World Meteorological Organization (WMO), which monitors global weather, stated:

(1) The first six weeks of 2014 have seen an unusual number of extremes of heat, cold and rain – not just in a few regions as expected in any winter, but right the way around the world with costly disruptions to transport, power systems and food production.

(2) "Melbourne, Adelaide, and Canberra have all had record heat waves, while temperatures in Moscow were 11C above normal.

(3) Germany and Spain were 2C above normal for January and this month has seen so far six major depressions develop over the Atlantic.

Unusual Heat Production is Producing Deadly Weather Events

Since 2014, because of the great heating of the earth, it brought unusual and deadly weather events such as volcanoes, earthquakes, tsunamis, hurricanes/typhoons, cyclones, floods, tornadoes, fires, snow, and other storm forces. The frequency and ferocity have changed to higher levels. Science states that they are getting worse in the presentations. Let us now review some of these storms and see how science projects their activities.

1. Strange and Unusual Occurrences of Firestorms

First, forest fires have increased worldwide. The United Nations (UN) warned that forest fires were at their worst in the history of fires. John Kielsen Gammon, a Texas A&M atmospheric scientist explains that the heat from the sun evaporates water from the soil or plants. All energy then goes into heating the ground and consequently heating the air. The heat will parch the forest making it even drier. The extra level of heat makes it possible for a fire to start (*National Geographic magazine*).

Second, data for firestorms 2017 to 2018 show that Queensland's ongoing firestorm number **110 fires,** whereas California has had some **7983 fires** (Andrew Glikson, 2018). The data show that the burning of the trees is increasing the CO2 levels and will make the effects of climate change even greater. The process is a continuous '360.'

Third, the National Aeronautics and Space Administration (NASA) states that the burning fires emit carbon dioxide. This and other greenhouse gasses moving among the land, air, and ocean affect climate change. Therefore, NASA concludes that the earth is in for a smoky, diseased, and fiery life if people do not make changes quickly (NASA.gov).

2. Strange and Unusual Occurrences of Earthquakes

First, the United States Geological Survey (USGS) research studies compared present earthquakes with average earthquakes since 1979. The findings state that more earthquakes are taking place than ever before. (The Extinction Protocol 2015; Conners 2012). The Earth and Planetary Science journal stated that the sun and its solar activity are triggering the frequency of earthquakes. Note:

1. Between January and April 2010, 27 earthquakes took place around the world
2. In 2011, an increase of magnitudes to 6.0 and 7.0 earthquakes almost topped the number in other years.
3. In 2012, 63 earthquakes took place in 12 months.
4. In 2013, 82 earthquakes with magnitudes ranging between 6+ and 8+ took place. (USGS)

Second, on August 14, 2019, the Swiss Center, and the European Mediterranean Seismological Center (CSEM- EMSC) information on earthquakes showed that in 7 hours and 5 minutes, some 50 earthquakes had taken place in various areas on earth. This means that if 150+ could take place in one day, therefore, 54,750 earthquakes may take place in one year. Sometimes, tsunamis will occur with earthquakes, especially in the sea. Statistics show that since 1771–the Great Yaeyama Tsunami, tsunamis have increased. From 1771 to 2000, there have been approximately 6 tsunamis that accompanied earthquakes. However, since the 2000s, there have been 7 tsunamis. Tsunamis are increasing along with the earthquakes (See Table No. 4 to find additional information on earthquakes and tsunamis).

Third, since the 1900s, seventeen major earthquakes were 7.0–7.9 and one great earthquake was 8.0 or above (USGS). A scientific company in the United Kingdom produced a graph showing how the magnitude of earthquakes has grown. It shows that the frequency and force of earthquakes seem to increase in recent years (see http://www.earth.webecs.co.uk/).

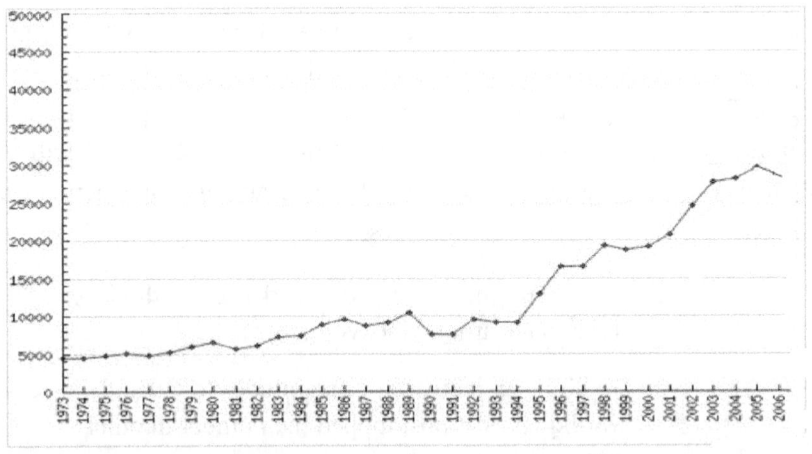

Chart showing the frequency and magnitude of earthquakes from 1973 - 2006

Fourth, findings on earthquakes show the occurrences of major earthquakes over one hundred years ago, based on data available from USGS (NEIC) and other online sources. The statistics show a drastic increase each year in the occurrences of earthquakes. Science identifies Global Warming and Climate Change as the cause for the increase.

Fifth, the US Geological Survey (USGS) stated that over 1280 earthquakes have hit Puerto Rico's southern area. Also, over two dozen of the earthquakes had a magnitude 4.5 or greater. A series of earthquakes occurred in Puerto Rico between December 2019 and January 13.

Presently, science is looking for answers for this series of earthquakes. They are concerned about the earthquakes' connection to the Puerto Rican Trench and the ocean tectonic plates of North America and the Caribbean, between which Puerto Rico lies. If Puerto Rico is having such activities, it means that these two tectonic plates of the earth are moving. Also, islands and areas of the mainland of North America could or may have problems with earthquakes or tsunamis. Also, scientists from NASA's Jet Propulsion Laboratory calculated a 99.9 percent probability that a 5.0 or greater earthquake will hit the Los Angeles region.

Strange and Unusual Occurrences of Hurricanes

First, according to the *Atlantic*, "Hurricanes really aren't supposed to get stronger than Hurricane Patricia. Hurricanes should not see sustained winds above 190 miles per hour in Earth's modern-day climate." Hurricane Patricia, which hit Mexico and Texas in 2015, had winds measured about 200 miles per hour (Meyer 2015).

Second, the Atmospheric Research (UCAR), stated that there has been observational evidence for an increase in intense tropical cyclone activity in the North Atlantic since about 1970. Also, science stated that sea- surface temperatures across the tropics have risen along with global temperature over the last hundred years. Also, they expect the earth to warm further in the next century. All else being equal, warmer oceans can support stronger hurricanes. The extra water vapor evaporated by oceans in a warmer climate will boost rainfall from hurricanes by as much as 8 percent for every 1.8°F of warming (Trenberth, 2014).

Third, the Weather Channel reported that the "2014 season featured the fewest number of named storms in seventeen years (eight storms), but also featured the strongest landfalling hurricane in the mainland U.S. in six years" The findings conclude that storms are becoming more active each year. In 2016, Hurricane Matthews was the most powerful storm of the 2016 Atlantic Hurricane Season. This hurricane hit South Carolina as a category 1 on October 8, 2016. In 2017, Hurricanes Irma (category 5) and Maria were among the devastating storms. In 2018, Hurricane Michael (category 4), slammed into the Florida Panhandle on October 10, 2018. In 2019, Hurricane Dorian (category 5) devastated two of the Bahamas islands, Great Abaco and Grand Bahama. These hurricanes brought much destruction, and sometimes the loss of lives.

Strange and Unusual Occurrences of Typhoons and Cyclones

First, typhoons and cyclones, which are like the Atlantic hurricanes, are now increasing in Europe and North America, where the central pressures of these storms are below 970 millibars (mb) and are causing a clear majority of devastation and loss of life. The scientific community predicts that the intensification of storms such as hurricanes, typhoons, and cyclones are all affected by global warming.

Second, the Atmospheric Research (UCAR), stated that there has been observational evidence for an increase in intense tropical cyclone activity in the North Atlantic since about 1970. Also, sea-surface temperatures across the tropics have risen along with global temperature over the last hundred years and are expected to warm further in the next century. All else being equal, warmer oceans can support stronger hurricanes. The extra water vapor evaporated by oceans in a warmer climate can be expected to boost rainfall from hurricanes by as much as 8 percent for every 1.8°F of warming (Trenberth, 2014).

Third, climatologists have published various analyses that correlate typhoon activity to climate change. Also, they have stated that such correlations show an increasing intensity of these storms as the earth's climate changes.

Fourth, the United Nations meteorological agency also found that climate change influences the severity of typhoons, such as Haiyan, which devastated the Philippines in 2013. The World Meteorological Organization stated that the record-breaking summers of high temperatures added to the increase in the global temperatures. This high peak of the global rising temperature caused the sea levels to rise, which worsened the situation in the Philippines when Haiyan hit. This problem causes storm surges to have a much more devastating effect than they had decades ago. The impact of Haiyan was significantly more than it would have been a hundred years ago because the sea level is higher. We may conclude that the storms are intensifying because of climate change, and the earth is getting hotter (Gearin, 2013).

Strange and Unusual Occurrences of Tornadoes

First, tornadoes affect several countries of the world. The data showed the United States, Canada, Russia, India, South Africa, Australia, New Zealand, Great Britain, and Argentina are targets of tornadoes. However, the central states of the USA have the highest tornado risk in the world. Scientists feel that the storms are changing in frequency and are wondering if a future warmer world will have more or fewer of these storms?" (Air Worldwide).

Second, in 2014, a study from Florida State University suggested that the strength and frequency of tornadoes hitting the United States had risen sharply since the 1950s. Professor James Elsner and his team examined historical weather data. The study outcome shows that the twisters that develop are stronger and increasing in frequency than ever. Also, when they come, "they come like there's no tomorrow." The study outcome suggests that tornado strength and frequency is increasing (Santa Maria 2014).

Third, a 2016 study on tornadoes concluded that extreme outbreaks have become more common. Also, the average number of tornadoes per outbreak has grown over 40 percent over the last century and the likelihood of extreme outbreaks is also greater (Tippett 2016). As noted, the studies showed that global climate change is the cause – the earth is warming.

Strange and Unusual Occurrences of Floods

First, the scientific community stated that weather changes cause massive flooding. Professor Peter Höppe, head of Munich Re's Geo Risks Research unit, stated: "weather conditions that lead to such flooding are becoming more frequent and that such weather systems remain stationary for longer" (Munich RE 2013). They believe that "the rise in emissions of greenhouse gasses and the production of heat causes the change in temperatures, therefore, have increased extreme (flooding) events.

Second, science states that Global Climate Change has caused massive and severe flooding in Germany (DW 2013), Australia (Fogarty 2011), and the United States. For example, in the United States between 1901 and 2000, the average rainfall was 2.04 inches. However, in 2009, October alone had a record of 4.15 inches. As a result, rivers overflow and flooding takes place. The National Center for Atmospheric Research (NCAR) stated that night thunderstorms could be a significant flash-flooding hazard through excessive rainfall (Davies 2015).

Third, scientists also state that severe flooding in Brazil is because of climate change. Professor Virgilio Viana of the Sustainable Amazonas Foundation said, "An unexpected change in weather means more water vapor and consequently more rain have stayed in the Amazon, causing river levels to rise."

Fourth, science concludes that there is a connection between flooding and climate change. Note:

(1) The rising temperatures have likely contributed to an increase in extreme rainfall events. As the earth's climate warms, the air can hold more water vapor, allowing for more intense downpours that can lead to flooding (Union of Concerned Scientists 2012).

(2) People are the ones that changed the landscape by removing the forest. There are no more leaves on the ground to absorb the water because they cut down the forest. Therefore, the water now runs freely off the ground into the rivers, and flooding is much easier" (Climatology Geographic Site 2015).

Strange and Unusual Occurrences of Meteorites

Since 2013 when the *Chelyabinsk region of Russia's Ural Mountains Meteor* hit, in 2018 another similar meteor exploded 15 miles away from earth over the *Bering Sea* (the National Aeronautics and Space Administration (NASA). It would appear that at this time there are not too many events of meteors. However, we do not know what will take place in the future (Molina, USA TODAY, 2019).

Concluding Thought

According to science, the phenomenon of Global Warming/Climate Change is real. The many weather-events show its reality. Also, it paints a dangerous picture for the future of humankind if such heating of the earth continues, as noted by scientists.

But looking at this information that science is sharing with the world, many people do not accept the concepts of science as real. Note the various beliefs. *First,* there are people who believe that the scientific phenomenon of Global Warming and Climate Change is only fake news. *Second,* there are people who say these events are *Mother Nature*. And she is having her way with events in the sky, on the land, in the waters, and with the weather. *Third,* there are people who do not know what to believe.

So, which one of the three views is right? The author believes that some people may have another view about the 'signs' of the strange and unusual events that are occurring in the sky, on the land, in the waters, and with the weather. Many people of the world, including the author, say, "there is another view, and it is the right one. The BIBLE gives the answers for the strange and unusual events that are taking place in the sky, on the land, in the waters, and with the weather." This view is so important, the author presented the information in *Three Sections and several short chapters* of the book to present this view of why strange and unusual events are taking place in our sky, our land, our waters, and our weather. What do you think the answer is? Read!

SECTION THREE

THE BIBLE'S OLD TESTAMENT PROPHETS PREDICT UNUSUAL 'END TIME' COMING SIGNS

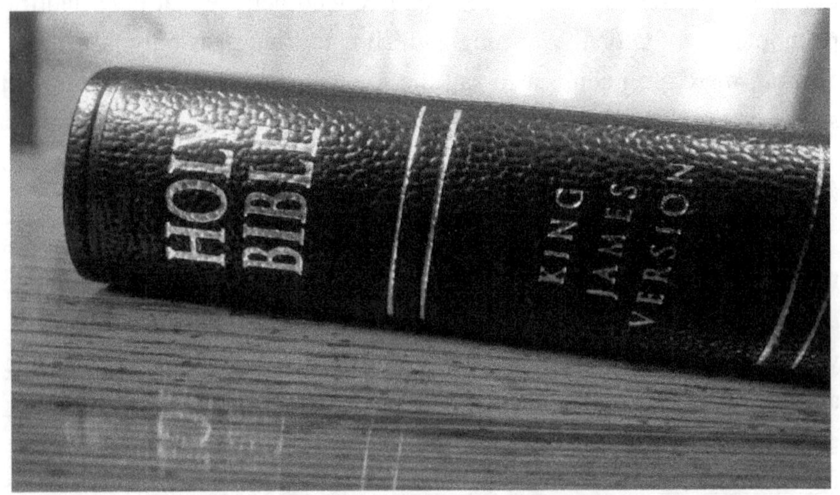

Background Information

The news media gave information on the stories of strange and unusual events occurring in the sky, on the land, in the waters, and with the weather. Also, science gave its views and beliefs about the occurrences of these strange and unusual events. Science has also given many predictions about the phenomenon of Global Warming and its Climate Change process. However, some people in the world do not believe the science of Global Warming and Climate Change. Of the people who do not believe in the science of global climate change, there is a group who believe that the events that are taking place are signs of the "End Time" or "Last Days." These are people who follow closely the prophetic teachings of the Bible, the Word of God. They feel that the Bible is the only book that has the real answers about the strange and unusual events that are occurring throughout the world. Therefore, with this background, let us now get some clarity on the Bible and its predictions about "End Time Signs."

First, the Bible is the inherent, Holy Word of God. There are two major writings of the Bible. The first portion of the Bible is THE OLD TESTAMENT and the second is THE NEW TESTAMENT. In the Old Testament, we find many prophecies or predictions by the Old Testament Prophets. The Bible states that God had unique ways of communicating with humankind. In the New Testament section of the Bible, we find many predictions or prophecies by Jesus Christ and the New Testament Apostles. The New Testament apostles, except for the apostle Paul, were first disciples of Jesus Christ during His ministry on earth.

Second, this book presents the prophetic messages of the Bible in *'three segments or sections.'* Each section presents predicted events that will take place in the sky, on the land, in the waters, and with the weather. The Old Testament Prophets prophesied their messages that God gave them about events of the 'end time'. In the New Testament, Jesus Christ prophesied His prophetic messages about coming events which the apostle Matthew classified as events of *'The End of the world'* (Matthew 24: 3). Also, in the New Testament, the Apostles of Jesus Christ prophesied the prophetic messages the Holy Spirit gave them about the *'End of the world.'* Together, the three *Sections* will explain all the prophetic messages concerning the *'Last Days'* or *'End Times'* as predicted in the Bible, God's Holy Word.

Third, in the *Book of Hebrews*, a book of the Bible, it states, **"God, who at sundry times and in divers manners spake in time past unto the fathers by the prophets, has in these *'last days'* spoken unto us by His Son, whom He has appointed heir of all things, by whom also He made the worlds"** (Hebrews 1:1-2). God used many prophets in the past. Many spoke of unusual events that took place in past times, some taking place today, and many to take place soon. The Bible refers to these events as prophecy. Many of the strange and unusual events that are taking place today in the sky, on land, in the waters, and with the weather, the prophets, apostles, and Jesus Christ predicted. Their predictions God must fulfill.

Fourth, people must understand that the Bible has embedded within its prophecies answers for unusual events, which it predicts as 'Signs of the End Time or *'End Times'*. Therefore, to give answers to the truths of the unusual events which are occurring right now in the world, this book explains in Sections Three, Four, and Five. For example, Section Three covers the prophetic predictions of the Old Testament prophets. Section Four covers the predictions of the Gospels of the New Testament (Matthew, Mark, Luke & John). Section Five covers the Epistles of the apostles Paul, Peter, and John. Also, each Section presents and explains the prophetic predictions in several chapters.

Fifth, with this idea in mind, this Section of the book now presents seven chapters of prophecies of the Old Testament prophets. However, before presenting the prophecies of each prophet, this Section gives information on the prophets and their prophecies.

The Old Testament Prophets and their Prophecies of Unusual Signs in the 'End Time'.

The Book of Hebrews states how in the *past*, God spoke to the people of that time through the ministry of the prophets. These prophets or *holy men of God* as the Bible calls them, shared prophecies. Many of the prophecies had events that will have signs in the sky, on land, in the waters, and with the weather. However, many of the prophecies were twofold. There was a message for the people of that day and also for people of our day. Also, many of the events had a close connection with the coming of the Messiah to the earth, Jesus Christ, of whom God spoke through "in the *'End Times'*.

With Biblical proof of how God communicates, let us now turn attention to His communication with the Old Testament prophets Isaiah, Ezekiel, Daniel, Hosea, Joel, Amos, and Zechariah. Each of these prophets gave prophetic events that will take place in the 'End Time.' This Section presents and discusses the prophecies of each prophet in seven short separate chapters. Let's read about their predictions to see how they relate to the phenomenon of Global Climate Change, if any.

Chapter 8

PROPHET ISAIAH PREDICTS UNUSUAL 'END TIME' SIGNS

The prophet Isaiah was one of the 'holy men' God selected to prophesy to the people of Judah and Israel.

Isaiah Predicts Unusual Signs in the Sky

God gave the prophet predictions of unusual events that will come to pass that relate to the 'sky' and the 'earth.' Isaiah's predictions about significant activities in the ***sky*** in these selected verses. Note:

(1) "For the stars of heaven and the constellations thereof shall not give their light: the sun shall be darkened in his going forth, and the moon shall not cause her light to shine" (Isaiah 13:10).
(2) "Then the moon shall be confounded, and the sun ashamed, when the Lord of hosts shall reign in mount Zion, and in Jerusalem, and before his ancients gloriously" (Isaiah 24:23).
(3) "And all the hosts of heaven shall be dissolved, and the heavens shall be rolled together as a scroll: and all their host shall fall down, as the leaf falls off from the vine, and as a falling fig from the fig tree" (Isaiah 34:4).

These are predictions about significant activities with the sun, the moon, the host of heaven, and the reigning of the Messiah on Mount Zion and Jerusalem. Such reign will be one that is glorious, for the glory of the Lord will be clear before his elders (Isaiah 24:23). Although these unusual events will take place in the ***sky***, they will also have a special manifestation on the earth. The selected scriptures are about unusual events related to Jesus Christ and his return to earth. *First*, he was to come as the Messiah for the Jewish people. *Second,* He will come as the bridegroom of his called-out assembly, the Church. *Third*, He will come back as, (1) *the true judge of humankind*, (2) *the King of kings, the Lord of lords,* (3) *the ruler of the millennial kingdom*, and (4), *the eternal light of the New Jerusalem and the new heavens and new earth.*

Isaiah's Predicts Unusual Signs on the Land, in the Waters, and with the Weather

Isaiah's prophecies of unusual events about waters and with the weather stated that **"And in that day, they shall roar against them like the roaring of the sea: and if one look unto the land, behold darkness and sorrow, and the light is darkened in the heavens thereof"** (Isaiah 5:30).

This prophecy states that there is a coming event in which the clouds will be so thick, that they will prevent the light of the sun, moon, and stars. Also, the stars will be affected. The event prophesied here says that when the Messiah comes back to the earth, His presence to the earth will cause the sun, the moon, and the stars to become darkened. However, before this prophecy occurs, another prophecy God shared with the prophet that must first occur. Note the prophecy:

> For unto us a child is born, unto us a son is given: and the government shall be upon his shoulder: and his name shall be called Wonderful, Counsellor, The mighty God, The everlasting Father, The Prince of Peace (vs. 5).
> Of the increase of his government and peace there shall be no end, upon the throne of David, and upon his kingdom, to order it, and to establish it with judgment and with justice from henceforth even forever. The zeal of the LORD of hosts will perform this (vs. 6-7).

First, between verses 5 and 6 of this prophecy, are prophetic events between the "child is born" and the "the son is given." However, many events will take place before the Son establishes His "government and the kingdom" on earth. For example, in Isaiah 53, God explained to the prophet how the message of the Messiah (1) people will not believe,

(2) they will not understand "the revelation of the arm of the LORD, " (3) the Messiah, Jesus Christ, will grow up, become despised of men, experience grief, and will receive wounds by being bruised by the people. Such treatment, Isaiah predicts, is for the transgressions, iniquities, and healing of all people (see Isaiah 53 for the full prophecy).

Second, among all this information God shared with the prophet that when the Messiah returns to set up his kingdom, not only will there be signs in the sky with the sun, moon, and stars, there will also be **"unusual sounds and noises from the roaring of the sea and the rushing of waters will distress people"** (Isaiah 17:12).

Regarding 'Signs of the End Time' the Bible predicts concerning the seas and waters, the science also predicts that global warming/climate change is producing signs today with the weather. For example, the weather is causing many unusual storms of hurricanes, tornadoes, cyclones, earthquakes, volcanoes, etc. as previously presented. As a result, there are many unusual signs with the roaring seas with tsunamis and other events. Does this mean that the prophet's prophecies are coming to pass, and the Messiah is soon to come? If this is the case, Isaiah's prophecy is partially fulfilling. Therefore, *Are people closer to these events than when Isaiah made his predictions of 'End Time' events?*

Let us now move on to the prophecies of the prophet Ezekiel.

Chapter 9

Prophet Ezekiel Predicts Unusual 'End Time' Signs

In the book of Ezekiel, the prophet prophesies about the signs of various events that will take place at the 'End Time'. Ezekiel's prophecies are about the Gentile world powers that will come against the Nation of Israel, with their armies, in the 'End Time'. When this event takes place, God will judge the nations. During the judgment of God on the Gentile nations, Ezekiel's prophecies state that there will be signs in the sky, on the land, in the waters, and with the weather. Note, **"And when I shall put thee out, I will cover the heaven, and make the stars thereof dark; I will cover the sun with a cloud, and the moon shall not give her light"** (Ezekiel 32:7).

The judgment that God shared with the prophet Ezekiel will have a significant effect on humans, animals, and other creatures that depend upon the light from the sun. The plants need light for photosynthesis or energy for survival. Humans need the sunlight for converting vitamin D. Solar energy also is a significant process today; this means that without the sunlight, there will be no power to take care of freezers, refrigerators, TVs, and all other appliances. Cell phones are one of the primary tools for communication today; they will have no power for operation. The world will experience great inconvenience. Also, a loss of power as previously described will affect humans' health. With presenting signs of this magnitude, people will be frightened, and some may also have nervous breakdowns, fainting, heart attacks, and even death. Let us now look closely into the events of the prophecy.

Ezekiel Predicts a Coalition of Nations to Make War Against the Nation of Israel in the 'End Time.'

First, this prophecy gives a few important events. Ezekiel identified a man by name, Gog. He will come from the land of Magog.

He will be the chief prince of Meshech and Tubal. God wanted this man to know that He, God, will judge

Second, God gives this coming *'leader from among the Gentile Nation'* a message of warning when he comes down into the land of Israel. The prophecy has not yet taken place but will have its fulfillment in the *'End Times'* (see Ezekiel 38:1–3).

Third, the prophecy describes three events. For example,

(1) Prince Gog of the land of Magog will lead a coalition of nations against the land of Israel.

(2) To form this coalition, God will bring this Gentile world leader to the land of Israel by the use of a strange method. Note, God will **"put hooks into thy jaws, and I will bring thee forth, and all thine army, horses and horsemen, all of them clothed with all sorts of armor, even a great company with bucklers and shields, all of them handling swords"** (Ezekiel 38:4). Whatever these "hooks in the jaws" are, God will hook them in the jaws and drag them down to fight against Israel. The Gentile nations will come with full armor. Ezekiel understood this 'full armor' to be swords, shields, spears, bows and arrows, and rock-throwers. These weapons were the best they had on that day. However, he would not have understood modern armies that now have helicopters, tanks, nuclear weapons, and troops that are well skilled and trained with these weapons of warfare (see Ezekiel 38:3-6).

(3) The prophetic texts in Ezekiel names some of the nations that will come against the land of Israel. 'Prince Gog' will be accompanied by, (a) Persia, (b) Ethiopia, (c) Libya, (d) Gomer, and all his descendants, (e) the people from the lineage of Beth Togarmah of the north. All these Gentile nations will come against Jerusalem and the land of Israel with a great confederate Gentile army (see Ezekiel 38:3–6).

Fourth, the prophecy states, **"Thus, saith the Lord God."** This is not 'fake news'. The events *will* take place. Jehovah God uses his personal name – LORD which means JEHOVAH. This great coalition of four excellent, well- trained nations and their allies will go to war against the land of Israel someday because God said it will happen. Keep in mind, such an operation cannot take place overnight. Therefore, it is possible that signs of the coming together of these armies are already taking place. As they progress, others will join the coalition for that final day. The Bible explains this event in the Book of Revelation. As the

Great Tribulation will end on the earth, "the Battle of that Great Day of God Almighty" will occur at a place named "Armageddon." As the battle begins, Jesus Christ will return to the earth at the Mount of Olives and will judge the nations during the battle (see Revelation 16:1-21& 19:11- 21).

Ezekiel Predicts that in the 'End Time,' the Gentile Nations Will Focus on the Land of Israel

God already said that Gog and the Gentile nations' confederation already have eyes on the land of Israel? Note what Ezekiel predicts in Ezekiel 38:14. It states, **"Therefore, son of man, prophesy and say unto Gog, Thus saith the Lord God; On that day when my people of Israel dwelleth safely, shalt thou not know it?"** Are great things taking place for Israel? Do they have safety in the land? God has much greatness for the nation of Israel.

Paul Alster (2015) stated in a Fox News article that there are potentially game-changing oil reserves discovered in Israel. The massive oil reserves in the Golan Heights, close to the country's border with Syria, is about 350 meters thick. Could the new-found oil be one of the *pruning hooks* that God will use to bring nations down against the nation of Israel? Michael Snyder (2015) asked this question in an Infowars article: "Will the discovery of huge amounts of oil in Israel lead to war in the Middle East?" He also stated, "The Israelis have discovered Billions of barrels of oil in Israel, and this discovery could make Israel energy independent for many decades to come."

Note the prediction of Ezekiel 38:15–16:

> And thou shalt come from thy place out of the north parts, thou, and many people with thee, all of them riding upon horses, a great company, and a mighty army: And thou shalt come up against my people of Israel, as a cloud to cover the land; it shall be in the latter days, and I will bring thee against my land, that the heathen may know me, when I shall be sanctified in thee, O Gog, before their eyes.

The coalition of nations and their armies will be so huge that they will completely cover the land of Israel. Important here, the Gog who will lead this coalition is coming from, **"thy place out of the north**

parts."

Ezekiel Predicts The Gentile Nations to Move Close to the Land of Israel in the 'End Time'

First, according to Ezekiel's prophecy, a nation from the far north will go down to the land of Israel. Could this nation be Russia? In fact, Russia has already moved troops into the nation of Syria, which is right next door to the land of Israel.

Second, the people of Israel are not living in peace. Many of the world's nations have a great hatred for them. Researchers have traced Gentile nations of the past and linked them to the existing nation today. For example, a publication by the Church of the Eternal God offers information on Middle Eastern and African nations in Bible prophecy. Note their classification of Bible nations with those that exist today:

Amalek (Esau's grandson): PLO and other violent groups
Ammon: Jordan
Aram, Arameans: Syria (with capital Damascus)
Asher: Belgium and Luxembourg
Assyria: Germany, Austria, German-speaking countries (small contingency in Iraq)
Benjamin: Norway and Iceland
Chaldea and Babylon: northern Italy (also southeastern France and parts of Spain and northern Africa)
Cush: Ethiopia, formerly Abyssinia (some descendants also in India and Sri Lanka)
Dan: Ireland and part of Denmark
Elamites: Iran, aka Persia (some descendants also in Poland) Esau
Ephraim: Australia, Canada, New Zealand, the United Kingdom, as well as part of South Africa Elam
Edom: Turkey (also referred to as Idumea, Teman, Bozrah, Seir)
Gad: Switzerland
Gomer: Mongolia *
Ishmael: Saudi Arabia (also referred to as Kedar, Hagar, Hagrites, Tema, Dedanites; his daughter Mahalath married Esau

(Edom)

Issachar: Finland Javan: Greece

Judah: the Jews, scattered among all nations, with a large number now living in the modern state of Israel

Kir: uncertain, Albania or Media or Moab or Egypt or Assyria or Babylon

Levi: scattered, some in Wales and Scotland

Lot: Jordan Lubim: perhaps part of Libya

Lydia: descendants of Egypt who settled in North Africa

Magog: China *

Manasseh: United States of America **Medes:** Russians and Ukrainians **Meshech:** Moscow, Russia

Moab: Jordan, perhaps also Western Iraq

Naphtali: Sweden Pathros: original land of Egypt, also India

Philistia: Palestinians

Put: Libya (or parts of Libya), also parts of northern and central India, Pakistan, and Bangladesh

Reuben: parts of modern France

Rosh: white Russians

Shinar: Iraq

Simeon: scattered, some in Scotland (especially in Glasgow)

Tiras, son of Japheth: American Indians

Togarmah: Siberia * **Tubal:** Tobolsk, Russia * **Zebulon:** the Netherlands

(http://www.eternalgod.org/booklet-2187/)

Third, from the listing above, maybe people can identify some of the countries in Ezekiel's prophecy who are predicted to come against the land of Israel. Besides the above listing of gentile nations by the *Church of the Eternal God,* the *Biblical Identity of Modern Nations* also has a listing that identifies Biblical nations and their modern names today. This listing states:

> Gog from the land of Magog is China.
> Meshech as Moscow, Russia Tubal is Tobolsk, Russia.
> Persia has not changed. It is Iran; "in present-day Iran are Persians" (Fyre 2009).

Put is Libya (or parts of Libya), also parts of Northern and Central India.

Cush is Ethiopia, formerly Abyssinia (some descendants also, in India and Sri Lanka).

Gomer is Mongolia. Togarmah is in Siberia.

Ezekiel Predicts Why God Will Allow the Gentile Nations to Come Against the Land of Israel

In Ezekiel 39:21–29, *first,* God gives a brief history of the nation of Israel and the Gentile nations. In this short history, he said that he would set his glory among these nations. God mentioned the reason for Israel's judgment. In this ruling, God stated that he turned his back on Israel because of their iniquity and unfaithfulness. He allowed them to go into captivity and placed them under the control of their enemies and even allowed some to die. God hid his face from the nation of Israel, and he wants the Gentile nations to understand this. It was not the power of the Gentile nations over Israel that caused them to be in captivity, for God allowed this to happen (see Ezekiel 39:22–23).

The *second* thing God did, He acknowledged that the people of the Nation of Israel paid for their transgressions over the years. They bore the shame for their unfaithfulness to him. Therefore, he wanted the Gentile nations to also know that the people who were once seen as captives, He will put them back into their promised land. He will have mercy on the house of Israel. He will gather them out of their enemies' lands. His act will allow many nations to see his holiness.

Third, when the nation of Israel understands and realizes what happened, they will know that the Lord their God, who sent them into captivity among the nations, has brought them back to their land. Also, he wanted them and the other countries to know that they would never go into captivity again, not even one. His face would always be toward them. As it was on the day of Pentecost when the Spirit came upon the disciples of Jesus Christ, so it will be during this time; the pouring out of God's Spirit will be in the house of Israel. The Lord God said this would happen as predicted by the prophet Hosea in a later section (Hosea chapter 2).

Ezekiel Predicts Unusual Signs Like Those of Global Climate Change in the 'End Time'

What science calls *global climate change* today will continue to manifest itself until the final day of God's judgment at Armageddon. On that day, as the coalition of nations with their armies come against the land of Israel and its people, God will intervene with many signs that are like today's events of global climate change.

Note what God told the prophet Ezekiel about the *'coming days.'* Note the prophecy:

> And it shall come to pass at the same time when Gog shall come against the land of Israel, saith the Lord God, that my fury shall come up in my face. For in my jealousy and in the fire of my wrath have I spoken.
> Surely on that day, there shall be a great shaking in the land of Israel. So that the fish of the sea, and the fowls of heaven, and the beasts of the field, and all creeping things that creep upon the earth, and all the men that are upon the face of the earth, shall shake at my presence, and the mountains shall be thrown down, and the steep places shall fall, and every wall shall fall to the ground.
> And I will call for a sword against him throughout all my mountains, saith the Lord God: every man's sword shall be against his brother. And I will plead against him with pestilence and with blood; and I will rain upon him, and upon his bands, and upon the many people that are with him, an overflowing rain, great hailstones, fire, and brimstone.
> Thus, will I magnify myself, and sanctify myself; and I will be known in the eyes of many nations, and they shall know that I am the Lord. (Ezekiel 38:18–23)

Ezekiel's prophecy predicts several signs will take place in the end time. The signs will be a magnification of the events science is attributing to global warming/climate change today. As the news media stories, there will be signs with the *fish of the sea*, the *birds of the heavens*, the *beasts of the field*, *all creeping things on the earth*, and *humans* who dwell on the earth. Rather than Global Climate Change, the Bible says, the sign will be "shaking at the presence of the Lord God." When we look at the animals and creeping things of the earth, the fires of Australia, the Amazon jungle, Africa, and California, it would appear that climate change is already causing the beginning of such judgment. But if the Bible is attributing these events to God. God is the One who is presenting "the signs" that science gives to climate change. Let us look closer at the signs of Ezekiel's prophecy.

A Great Earthquake, Deaths of Fish, Birds, Animals, and People in the 'End Times'

First, there will be a ***great earthquake*** in the land of Israel. This earthquake will be because of God's fury. The earthquake will be so huge that the Richter scale cannot measure such an earthquake. To further understand what we are describing, we should understand the scale which measures earthquakes. Charles F. Richter of the California Institute of Technology, in 1935, developed this scale as a mathematical device to compare the size of earthquakes (United States Geological Survey (USGS)– Science for a Changing World). This is the measurement used but in this case of Ezekiel's prophecy, the scale just will not work. Ezekiel's prediction of this earthquake will be so great that the mountains will overturn, the cliffs will crumble, and every wall will fall to the ground. Besides what Ezekiel predicts, the prophet Zechariah and the apostle John also made similar predictions about this coming great earthquake. It is further presented and described in chapters dealing with their prophecies.

Second, God will pour down ***torrents of rain***. There will be floods from this rainfall. Hurricanes carry torrents of rain. However, in this case, God will intensify the downpours of rain but will also add to the downpours, ***burning sulfur*** to fall from the skies on Gog and his troops and the many nations.

Third, ***hailstones and burning sulfur*** will fall from the skies on Gog and his troops and the many nations with him. Normally, hailstones fall with an icy appearance. But in this prophecy, the hailstones that will fall upon the earth will have burning sulfur, thus, burning sulfuric hailstones. What a storm this will be.

Fourth, ***disease and plague*** will be present on the land and among the dwellers of the armies. Global warming/climate change is in its greatest display. Other scriptures also predict such events of plagues and diseases to take place on the earth.

Fifth, the animals of the sky, earth and sea will die. "The ***fish in the sea***, the ***birds in the sky***, the ***beasts of the field***, ***every creature that moves along the ground***, and ***all the people on the face of the earth will tremble at my presence."*** Such events are already taking place in the sky, on the earth, in the seas. Science states that these kinds of unusual events are happening because of global climate change. However, here they are prophetic signs of the "End Time."

Sixth, signs will confuse people so much that, "***Every man's sword will be against his brother."*** The soldiers will kill their fellow soldiers. The interesting point here, God does not have to use weapons such as those used by humans. He uses the natural things around us and amplifies them in a way that humankind is overcome by them.

The Results of the judgment on the Gentile Coalition will produce events like those science labeled as Global warming/climate change. Like the unusual events that are taking place today, on that day, there will be strange and unusual events. There will be: (1) unusual deaths of birds, fish, and animals; (2) strange and unusual patterns of the weather; (3) strange and unusual events or signs will happen in the sky, on land, and in the waters; (4) God will come from heaven and intervene on earth on behalf of the people of the land of Israel. This last sign is very unusual, for, God will come from heaven to the earth. Can you imagine the response of the people of earth when they behold God?

Ezekiel's prophetic predictions will manifest strange and unusual signs like the events/signs science believes to be outcomes of global climate change. However, they will come with greater magnitude and velocity. Other prophets and apostles also shared this prophetic message that Ezekiel received from God. Note, when events listed in the Bible occur more than once, we know it as "the law of double reference." The law of double reference here means that God will do what He says He will do. God today is showing that He still loves humankind and is giving various strange and unusual signs in the sky, on land, in the waters, and with the weather to get people's attention. He is allowing such signs so that humankind will take note and come to him.

The prophet Daniel has also prophesied about similar events like those of Ezekiel to take place in the *'End Times'*. Signs he predicts to occur in the sky, on the land, in the waters, and with the weather. Let us now read about the prophet Daniel "End Time Signs." You will be surprised by what signs Daniel is predicting for the "End Times."

Chapter 10

Prophet Daniel Predicts Unusual 'End Time' Signs

Nebuchadnezzar's army took Daniel into captivity at a young age (Daniel 1:1–4). Along with Daniel were three other young men—Hananiah, Mishael, and Azariah. To these four young men, the prince of the eunuchs gave them new names. He gave Daniel the name of Belteshazzar; Hananiah became Shadrach; Mishael became Meshach; Azariah became Abednego. Of these four young men, Nebuchadnezzar was favorable to Daniel. He allowed him to serve in prominent positions in his government.

Daniel was a faithful man of God who prayed several times a day. In his daily prayers, he never forgot his homeland of Judah. In Daniel's captivity in Babylon, God called him to be His servant to the Nebuchadnezzar, and his grandson, Belshazzar. Then later to Darius and Cyrus, leaders of the Medo- Persian Empire.

Daniel was faithful in this practice, even in the land of Babylon where their laws stopped people from praying to a different god than theirs. To disobey this law, the penalty was death. However, people spied on Daniel. The king had him cast him into the den of lions. God protected by sending an angel who shut the mouths of the lions (Daniel 6:22).

After this experience of Daniel with the lions, God gave Daniel many dreams and visions about the future of the Israeli people, the nations of the Gentiles and about events like those of present global climate change, that would take place in the '*End Time'*. Some of the prophetic visions that Daniel the prophet received from God had great significance with the sky, the land, the waters, and the weather. Among the prophetic events Daniel received from God, some were about the '*End Time'* increase of knowledge and a blueprint of the 'End Time' events. Let us now investigate Daniel's prophecies about important signs or events that would take place in the 'End Time.'

Daniel Predicts Strange and Unusual Signs Among the Gentile Nations

The First Prophecy being investigated is Daniel 2:31–45. This prophetic text is concerning a dream King Nebuchadnezzar of Babylon had. The dream troubled the king. He sought someone to relate to him his dream and the interpretation. After none of the king's interpreters could tell and interpret his dream, someone told the king about Daniel and his power to interpret dreams. The king summons Daniel to present and interpret his dream. Daniel asked the king for time to seek God for the answer. God gave Daniel the dream and its interpretation and Daniel gave it to the king. Note the dream and its interpretation.

The Dream of King Nebuchadnezzar of Babylon

1. God revealed to King Nebuchadnezzar in a dream, a great image that was huge and gave off a brilliance
2. The image had a head of fine gold
3. The image breast and arms looked like silver
4. The belly and thighs of the image looked like brass
5. The legs and feet of the image was part of iron and part of clay
6. A stone cut out without hands, it smote the feet of iron and clay of the image and broke them in pieces.
7. The destruction of the feet of the image by the stone caused (a) the legs of iron and clay, (b) the belly and thighs of brass, (c) the breast and arms of silver, and (d), the head of gold to all broke into pieces, and (e) the broken pieces of the image became like little particles which the wind carried away.
8. The stone which smote and destroyed the image became a great mountain, which filled the whole earth.

Interpreting King Nebuchadnezzar's Dream(Daniel 2:31–45).

The prophetic scripture references the entire Gentile kingdoms and nations. More than being a *sign of the times*, this is a historical representation of the Gentile nations' power of rulership on the earth. The image of a man that King Nebuchadnezzar saw in his dream depicted all the ages and ruling kingdoms of the Gentiles. The Gentile-Kingdom reign started with the Kingdom of Babylon until the Second Coming of Christ Jesus (the destroying Rock, establishing the eternal kingdom of God). To expound on the history of the Gentile nations rule, (1) it started with Nebuchadnezzar and Babylon, (2) followed next by kingdom of Medea Persia, which today is Iran and Iraq, (3) the nation that followed Media Persia was Greece, (4) the Roman Empire was the fourth kingdom God established after the kingdom of Greece.

The Roman Empire is the last kingdom God gave to the Gentile nations. Also, no nation ever conquered the fourth kingdom – the Roman Empire. It fell apart by internal derogation. This kingdom has become fragmented, and it will again come to power in the "End Times." It will come to power again as the 'Fragmented Roman Empire.' This fragmented Roman Empire is today operating with each nation's agenda. As the world looks for a leader from the Fragmented Roman Empire, this leader will emerge with full endowment of the powers of Satan. He will come into power and will be the world dictator, the Antichrist, and Beast as the apostle John beholds him, and the Little Horn and Prince as the prophet Daniel saw him. So, note Daniel's prophecy of the Gentile nations and the "Signs of the End Time."

Daniel's Vision of the Sign of the Gentile Nations Coming to an End on Earth

The Second Prophecy being investigated is Daniel 7:2–14. This second prophecy predicts the destruction of the Gentile world power as seen by the prophet Daniel in this vision God gave him. Note the prophecy:

> ***Part 1:*** *God Revealed to Daniel the 'Emerging Fragmented Roman Empire' from among the Gentile Nations*

Note Part 1 of the Second Prophecy:

Daniel spoke and said, I saw in my vision by night, and behold, the four winds of heaven strove upon the great sea. And four great beasts came up from the sea, diverse one from another. The first was like a lion that had its wings plucked, and it was lifted from the earth and made to stand upon the feet as a man, and a man's heart was given to it. And behold another beast, a second, like to a bear, and it raised up itself on one side, and it had three ribs in the mouth of it between the teeth of it: and they said thus unto it, Arise, devour much flesh. After this I beheld, and lo another, like a leopard, which had upon the back of it four wings of a fowl; the beast had also four heads, and dominion was given to it.

After this, I saw in the night visions, and behold a fourth beast, dreadful and terrible, and strong exceedingly; and it had great iron teeth: it devoured and break into pieces, and stamped the residue with the feet of it: and it was diverse from all the beasts that were before it, and it had ten horns. I considered the horns, and behold, there came up among them another little horn, before whom there were three of the first horns plucked up by the roots: and behold, in this horn were eyes like the eyes of man, and a mouth speaking great things (Daniel 7:2-8).

*Interpreting **Part 1** of Daniel's Second Prophecy – the Vision*

In examining this prophecy, Daniel again saw the four Gentile kingdoms which God revealed to him in Nebuchadnezzar's dream. However, instead of the kingdoms being represented in four *metals* of gold, silver, bronze, and iron, God presents them to Daniel as *beasts,* lion, bear, leopard, and an unknown beast. One who appeared to be part of the first three beasts. It had teeth like a lion, feet like a bear to stump and destroy, and it was fast like a leopard in action.

First, he saw "the four winds of heaven strove upon the great sea." The "four winds of heaven" illustrates the power of God. It shows that all the events of the earth, God knows and is in control of them.

Second, out of this vast sea, he saw "four great beasts coming from this great sea, each different from the other." The 'sea' represents the world and its diversity of people, powers, and leadership. The "four great beasts" who came from the sea, came from the world of humankind.

Third, note the diversity of the Gentile world powers.

The *first beast* *was like a lion with wings that resembled those of the eagle.* This kingdom was that of Nebuchadnezzar. The lion is the symbol of Babylon.

The *second beast* *that came out of the sea resembled a lopsided bear.* This beast had three ribs in his mouth, held with his teeth. The creature received the command to devour a lot of flesh. The *lopsided* view of this kingdom represents how the kingdom of the Medes and Persians were always divided and had conflict. Each of the nations within this kingdom rule one time or another. Even today they are still at odds with each other. However, they were conquered by Alexander the Great.

The third beast resembled a leopard that had four wings attached to its back that resembled wings of a fowl and attached to its body were four heads. God gave this 'Gentile' leader/beast – Alexander the Great, much power. History states how Alexander the Great came and conquered the known world with speed. He exalted his kingdom, but it soon fell apart after he died around the age of 33 years. His four generals divided the kingdom among themselves. Therefore, this beast represents the leader and the Greco-Macedonian Gentile Kingdom. The kingdom was defeated by the Roman Empire.

The fourth beast that came up out of the sea is *"dreadful and terrible, and strong exceedingly."* Note the activity of this *fourth beast*:

- The *fourth beast* had iron teeth and used them to break into pieces that were in its way and stamped the residue of the broken pieces with its feet. This beast was entirely different from the other creatures.
- This *beast* had ten horns.
- From the ten horns came another beast named the *"Little Horn"*
- The *"Little Horn"* is (1) powerful, (2) He has eyes like a man, (3) He will come speaking and convincing the masses, (4) He will destroy three of the ten kingdoms on the earth, leaving seven 'horns' of power under him.

All of the above four 'Beasts' are representatives of the Gentile Kingdoms God gave to them. God revealed to Daniel the history of the Gentile world powers. The first Beast represents Babylon. The second

Beast represents the Medes and the Persians. The third beast represents the Graeco- Macedonian Empire, and the fourth beast represents the Roman Empire. Each empire came to power, ruled for such a period and the less inferior conquered the former stronger. However, no nation ever conquered the Roman Empire. It fell apart by internal derogation. This power remains in the world operating in a fragmented way. One day, from among the Fragmented Roman Empire and its Religious System, the 'Little Horn' will come forth. He will be the Antichrist which the apostle John predicts in the Book of Revelation. Yet, he will appear as a *religious man*. He will be the worst dictator the world will ever experience. The Antichrist, with the assistance of his associate the False Prophet, will make the people of the world worship him as a god.

Part 2: *God Revealed to Daniel the Defeat of the Fragmented Roman Empire's Powers and its Beast*

Note Part 2 of the prophecy:

I beheld till the thrones were cast down, and the Ancient of days did sit, whose garment was white as snow, and the hair of his head like pure wool: his throne was like the fiery flame, and his wheels as burning fire. A fiery stream issued and came forth from before him: thousand thousands ministered unto him, and ten thousand times ten thousand stood before him: the judgment was set, and the books were opened. I beheld then, because of the voice of the great words which the horn spake: I beheld even till the beast was slain, and his body destroyed and given to the burning flame. As concerning the rest of the beasts, they had their dominion taken away: yet their lives were prolonged for a season and time (Daniel 7:9-12)

*Interpreting **Part 2** of Daniel's Vision*

In unpacking the above section of Daniel's prophecy, several events are to take place when and after the Fragmented Roman Empire is set up.

First, all the thrones, kingdoms, or world powers of the Gentile rulers will end. He saw the beginning of the Gentile world powers and their behavior within politics and power until God, The Ancient of Days destroys them.

Second, The Ancient of Days which refers to God, will set up His Kingdom on the earth and will sit on His Throne. The Bible describes "The Ancient of Days" as One who wore a garment that was "white as snow, and the hair of his head like pure wool." His throne is a fiery flame, and his wheels are burning fire. Also, from him came forth a stream of fire, and before him stood thousands, and thousands ministered unto him as they stood before him.

Ezekiel also saw '*The Ancient of Days'* in his vision in chapter 1 in all His glory down by the River Chebar (Ezekiel 1:1-28). Also, the prophet Isaiah saw Him in a vision, the year when King Uzziah died, in all His glory(Isaiah 6:1-5). In all cases, *The Ancient of Days* appeared in His glory.

Third, Daniel saw that at the end of all the political rhetoric, misinformation, misunderstandings, and lies by the world leaders, God, *The Ancient of Days,* will judge. Therefore, it is good to know Jesus Christ because of all judgments God has committed to the Son, Jesus Christ. He is the express image of God, the One who made all things (Hebrews Chapter one). He is "God," He will judge. Therefore, we can say, 'The Ancient of Days' is Jesus Christ. Here again, the scripture shows the uniqueness of the Godhead. All three persons are co-equal, co-efficient, and co-eternal. To even demonstrate this truth, the apostle John while in exile on the Island of Patmos, saw Him, Jesus Christ in His glorified form. His head and hair were white like wool and snow. His eyes looked like flames of fire. His feet were like fine brass and his voice sounded like great rushing water of many streams, rivers, and waterfalls. His appearance was like the shining sun (Revelation 1:14-16).

Fourth, the prophetic scriptures show that God will judge the Gentiles out of the records written in the "Books" – they will be opened.

Fifth, the Beast, Antichrist, or the 'Little Horn' God will judge and cast him into the burning flame.

Sixth, in Daniel 7:12, it states that God will prolong the other 'seven horns' powers/lives for a "season and time" or until God is ready to judge. In one season, Satan will deceive and will lead them in a war against Jesus Christ the Messiah, his saints, and the holy kingdom that the Messiah will set up on earth. John the apostle, many years after the prophet Daniel, saw this same vision but in greater detail, the book of

Revelation records. As noted by the apostle John in the book of Revelation, Daniel's vision that God gave him allowed him to view the coming of the Messiah to the earth. The apostle John saw in his vision where God will open the books and He will judge them out of the things written in the books. Whoever names God has not written in the Lamb's Book of Life, at the Great White Throne, God will judge them and cast them into the burning fire (see Revelation chapter 20).

All these events are further explained in the New Testament prophecies, especially those of the apostle John which in the Book of Revelation. However, let us continue with this prophecy of the prophet Daniel.

Part 3: *God Revealed to Daniel the Coming of the Son of Man and the Establishment of His Kingdom*

Note *Part 3* of the prophecy:

> I saw in the night visions, and behold, one like the Son of man came with the clouds of heaven and came to the Ancient of Days, and they brought him near before him. And there was given him dominion, and glory, and a kingdom, that all people, nations, and languages should serve him: his dominion is an everlasting dominion, which shall not pass away, and his kingdom that which shall not be destroyed. (Daniel 7:13-14).

Interpreting Part 3 of Daniel's Vision

The vision grieved Daniel. He could not comprehend what was going on. *First*, Daniel saw the Son of man coming in the clouds of heaven. *Second,* he saw Him going to the 'Ancient of Days – God. *Third,* Daniel saw the Ancient of Days – God, giving to the Son of Man, (1) dominion, (2) Glory, and (3), a Kingdom. *Fourth*, all people, nations, and languages will serve him. *Fifth,* the Kingdom which God will give to the Son of Man, will be everlasting.

Note, Daniel had not seen the book of Revelation where Jesus Christ shared with the apostle John the many strange and unusual 'Signs of the End Time." Yet, God gave him a vision that would take place in the future among the nations in the end times. However, this event and many other events God gave to the apostles of Jesus Christ, especially John, about these future events. Today the full revelation of the events that God has in his program is in the book of Revelation, including some which are like the visions that God gave Daniel. Some have already taken place, and others are to come to pass.

Daniel's Predictions of Unusual Signs Regarding the Gentile World Powers and the Nation of Israel

Here in Daniel's visions, we see activities described that refer to the sea, the land, and the sky. However, regarding the heavens, we see that the Son of Man will come with the clouds of heaven and his saints. This event is the Second Coming of the Messiah, which has not taken place yet. People will see all these activities on the land throughout the world with the Gentile nations. Remember the history of the kingdoms before. Nebuchadnezzar led the Babylonian Empire. His grandson Belshazzar next ruled. However, Medes and Persians (Iran and Iraq) defeated the grandson.

The next power was that of the Greek-Macedonian Empire, led by Alexander the Great, and finally the fourth kingdom, the Roman Empire. As mentioned, no nation ever conquered the Roman Empire but fell apart, leaving behind the fragmented Roman Empire. We know that there was a ten-nation alliance made up of Belgium, France, Germany, Greece, Italy, Luxembourg, Netherlands, Portugal, Spain, and the United Kingdom (a fulfilled prophecy). Are these the nations that Daniel saw in his vision? Whether or not they are all Gentile world powers and their kingdoms, God will take them away and give them to the saints of the highest God, according to the book of Daniel.

Strange and unusual activities will continue in the sky, on land, and in the waters. The Israeli people rejected the rule of theocracy. Instead, they asked God for a king to rule them as the other nations had. But Daniel's prophecy states that God will institute theocracy once again, forever. God will again deal with his people, of whom Daniel was prophesying. However, God gave Daniel a prophetic promise to remind all people that those who believed in God and had been following him, even though they died, had a future promise of life. However, those who did not follow Him would live in misery. Notice what the prophecy of

Daniel 12:2 says, "And many of them that sleep in the earth's dust shall awake, some to everlasting life, and some to shame and everlasting contempt" (emphasis added).

Daniel Predicts Unusual Signs to Come in the "End Time"

Approaching the close of Daniel's prophetic ministry, God gave the prophet a message about **the Last Days or the Time of the End**, and the *signs* to look out for. Also, the evidence of these signs will be proof that the "time of the end' is near. In a conversation between God and Daniel, God said, **"But thou, O Daniel, shut up the words, and seal the book, even to the time of the end: many shall run to and fro, and knowledge shall be increased"** (Daniel 12:4). This message which God gave Daniel is in the Book of Daniel in the Bible, and it is one of Daniel's prophecies that has and continues to come to pass.

Note the Fulfillment of Daniel's Prophecy

In considering the advancements of the centuries, we have much evidence that knowledge among humankind has rapidly increased. For example, there have been benefits of education across the globe. Technologies have gone into the entire world in mathematics and medicine. Devices have sprung up that have enabled all these innovations—telephones, airplanes, computers, power plants, automobiles, generators, motors, and more. All around the world there are technologies for mining minerals, oil, iron ore, and engineers are developing and upgrading all the time. Factories with new operations are in place around the world. Production rates within the factories are faster. Yes, there is proof that knowledge has increased and is still increasing.

The National Academy of Engineering (2007) has a listing of the twentieth-century benefits to society over one hundred years. Note the list of twenty innovations:

> 1. electrification, 2. Automobile, 3. Airplane, 4. water supply and distribution, 5. Electronics, 6. radio and television, 7. agricultural mechanization, 8. Computers, 9. Telephone, 10. air conditioning/refrigeration, 11. interstate highways, 12. space flight, 13. Internet, 14. Imaging, 15. household appliances, 16.

health technologies, 17. petrochemical technology, 18. laser and fiber optics, 19. nuclear technologies, and 20, high-performance materials (National Academy of Engineering, 2007).

Wikipedia stated there had been a "scientific revolution was the emergence of modern science during the early modern period … developments in mathematics, physics, astronomy, biology, and chemistry … began in Europe … and continued through the eighteenth century."

In the article "The Greatest Century That Ever Was: 25 Miraculous Trends of the Past 100 Years," Julian L. Simon and Stephen Moore stated:

> there has been more improvement in the human condition … Gigantic strides have been made in living standards in most other parts of the world … Human Life Span, Death Rate of Children and Mothers, Infectious Diseases, Cancer and Heart Disease, National Output, Wages, Wealth, Poverty, The Workweek, Farm Productivity, Entertainment, Housing, Electrification, Communications, Transportation, Inventions, The Information Age, Computers and the Internet, Education, Accidental Deaths, Environmental Quality, Natural Resources, Sexual Equality, Racial Equality, Conclusion: The Greatest Resource. (Libertarianism.org. 1999)

Because of the achievement of technology, especially computer technology, the entire world is a different place. Today people can contact others all around the world in seconds. These technologies are in servers, desktops, laptops, notebooks, tablets, smartphones, and so on. Computer companies are developing capabilities by the minute. Each year the phone companies are putting out smartphones with better technology. Today the smartphone can do about anything, short of taking a person to the bathroom. The MIT Technology Review talks about the "Ten Breakthrough Technologies for 2017." This is what the Review had to say:

1. These technologies all have staying power. They will affect the economy and our politics, improve medicine, or

influence our culture. Some are unfolding now; others will take a decade or more to develop. But you should know about all of them right now. These technologies include 'Reversing Paralysis' - Scientists are making progress at using brain implants to restore the freedom of movement that spinal cord injuries take away.

2. Self-Driving Trucks – Tractor-trailers without a human at the wheel will soon barrel onto highways near you. What will this mean for the nation's 1.7 million truck drivers?
3. Paying with Your Face – Face-detecting systems in China now authorize payments, provide access to facilities, and track down criminals. Will other countries follow?
4. Practical Quantum Computers – Advances at Google, Intel, and several research groups show that computers with previously unimaginable power are finally within reach.
5. The 360-Degree Selfie - Inexpensive cameras that make spherical images are opening a new era in photography and changing the way people share stories.
6. Hot Solar Cells – By converting heat to focused beams of light, a new solar device could create cheap and continuous power.
7. Gene Therapy 2.0 – Scientists have solved fundamental problems that were holding back cures for rare hereditary disorders. Next, we'll see if the same approach can take on cancer, heart disease, and other common illnesses.
8. The Cell Atlas – Biology's next mega-project will find out what we're made of.
9. Botnets of Things – The relentless push to add connectivity to home gadgets is creating dangerous side effects that figure to get even worse.
10. Reinforcement Learning – By experimenting, computers are figuring out how to do things that no programmer could teach them. (MIT Technology Review 2017)

Remember, the three areas of the prophecies of the prophet Daniel that will take place *in the sky, on land,* and especially in the *'End Time'*. *First,* the predictions about the events of the Gentile nations and their activities on earth. These have begun and will continue until the return of Jesus Christ to the earth. *Second*, the predictions about the signs of the Gentile nations coming against the Nation of Israel. *Third,* the increased knowledge of technologies Daniel predicted will assist in the wars and the world's nations' greed for them.

The above prophecy Daniel predicted is being fulfilled. So, don't you think that the others about the "Little Horn" or "Antichrist" God will also fulfill? Don't you think The Ancient of Days will judge the Gentile nations? Don't you think in the end, God will resurrect people who died? He will give eternal life to some and others shame. Yes, all Daniel's prophecies God will fulfill. Therefore, look for the coming of strange and unusual signs. They will appear in the sky, on the land, in the waters, and with the weather to occur, just as Daniel predicts.

Chapter 11

Prophet Hosea Predicts Unusual 'End Time' Signs

Among all the increase of knowledge and communication Daniel prophesied about, it seems that humankind is still ignorant as to many of the strange signs or events in these *'End Time'*. Hosea the prophet admonishes us to,

> Hear the word of the Lord, ye children of Israel: for the Lord hath a controversy with the inhabitants of the land because there is no truth, nor mercy, nor knowledge of God in the land. By swearing, and lying, and killing, and stealing, and committing adultery, they break out, and blood toucheth blood. Therefore, the land will mourn, and every person who lives on the land will languish. Also, the beasts of the field, the fowls of the heaven, and also, the fish of the sea shall be taken away. (Hosea 4:1–3)

These are short passages of biblical predictions, but they influence the sky, the land, and the waters. *First*, with the people of the land, whose lives are a projection of (1) swearing or profanity, (2) no truth but lies, (3) killing, (4) stealing, and (5) acts of adultery. This is the way God saw the people of that time. As a result, the people will be (1) a mourning in the land because people will suffer, (2) death to the beast of the field – the animals will die, (3) the fowls or birds of heaven will also die, and (4), so will the fish of the sea.

Interpreting Hosea's prophecy is twofold. It affects the people of the land and also, the animals of the sky, land, and sea. Therefore, let us investigate the findings of these prophetic scriptures. They appear to be like the signs we see today.

Hosea Predicts the Sign of Unusual Condition of People in the 'End Time'

In the days in which the prophet Hosea lived, God reminded him of the vision he gave him—that the land was devoid of mercy, there was a lack of truth among them, and there was a decreasing knowledge of God among them. Thus, there would be certain unusual signs or events that would take place, not only in the sky but also on land and in the waters.

Hosea Predicts an Increase in Murders in the 'End Time'

People will have no mercy toward each other in the *End Time.* According to the prophecy, people will express hate and cruelty throughout the world. Are there cruel people on the earth today? Look around. Are there acts today where people shed innocent blood?

First, in answering these questions, throughout the world people seem to lack *mercy* for each other. At home and abroad, an individual or group will take off a person's head for no reason at all. Many times, among people who claimed to be *religious people*, they engage in merciless acts of this sort in the name of some god. Others may not kill another because ofreligion, but they kill from within their hearts because of stored up hate they carry within.

Second, in areas of the world, many wars are ongoing. These wars kill many innocent people, including children. People are constantly fleeing for refuge. But the governments that wage such wars seem not to care. Even the innocent people who are killed, they call the killing *collateral damage* as if these people are animals. Sorry, but they were just in the fighting's way and were killed.

Third, in the United States, people are killing each other with guns, knives, and other methods. In some areas, the youth are engaged in territorial drug wars. As a result, many of those who are engaged in these drug wars lose their lives. Also, sometimes there are innocent people who are also killed by stray bullets from the guns of these people. Yet, they do not care, because the money from drugs is more important than the lives of people. Then there are the hate wars because of diversity in the land. Some people feel that others are taking from them their space, authority, and livelihood. Hatred and racial reasons appear to be the underpinnings for deeds of violence. On and on we can describe the hate, bloodshed, and loss of life within the United States. The picture is like that of the prophet Hosea.

Hosea Predicts Dishonesty Among People in the 'End Time'

God told Hosea that he was bringing judgment because there is *no truth* in people. To have no truth in one's life is to tell a lie. Today, people just lie outright without thinking. People lie to achieve a position, a job, or to escape a judgment that may befall them. Politics today is full of lies. In fact, in the election process in America, each candidate calls the other a liar. They accuse each other of lying about their records and what they promise the people. No wonder, in the book of Romans, the apostle Paul mentioned, "Not at all! Let God be true, and every human being a liar" (Romans 3:4). Also, the apostle John stated, "all liars God will judge" (Revelation 21:6). Could it be that the lack of mercy is because people have no knowledge of God in them? Well, God told Hosea about one additional sign of the *'End Time'* people will display. That sign will be "the lack of knowledge of Him."

Hosea Predicts A Decreasing Knowledge of God Among People in the 'End Time'

First, God told Hosea that the condition of the land has people suffering. Therefore, He will take away the birds from the sky, the animals from the land, and the fish from the sea. The people were not thankful for what God had given them. Therefore, they went away from God. Their knowledge of Him became limited. Their worship of him was no longer in 'Spirit and in Truth.' In fact, the people in Hosea's day became corrupted. It was so bad, God told Hosea to marry a harlot. Also, they will call their son 'Loammi', which means, "for you are not my people, and I will not be your God (Hosea 1:9).

Second, the prophecies of Hosea agree with the predictions of the apostle Paul concerning how people will become in the *'End Time'*. For example, the apostle Paul predicted:

> Every person must know that in the *'End Time'* perilous times shall come. For men shall be lovers of their own selves, covetous, boasters, proud, blasphemers, disobedient to parents, unthankful, unholy, Without natural affection, truce-breakers, false accusers, incontinent fierce, despisers of those that are good, Traitors,

> heady, high-minded, lovers of pleasures more than lovers of God; Having a form of godliness, but denying the power thereof: from such turn away. (2 Timothy 3:1–5)

Third, God gave this prophet many object lessons to get the attention of the people. Yet, they remained without *mercy*, *truth*, and *knowledge of the Almighty God.* As a result, there were signs in the sky, on the land, in the waters with birds, animals, fish, and people. The prophetic lesson resembles the signs of the *'End Time'* that other prophecies predict.

Hosea Predicts the Sign of Unusual Mass Deaths of Animals in the 'End Time'

> "the beasts of the field, and with the fowls of heaven; yea, the fishes of the sea also shall be taken away."

First, this portion of Hosea's prophecy lends itself to the predictions of many other prophecies of the prophets, Jesus Christ, and His apostles. Also, according to the findings of the News Media stories regarding Global Warming and Climate Change, the prophecies seem to go along with many of the findings of the massive deaths of animals in the sky, land, and sea, in the world today. God did not tell Hosea how the animals will die.

Second, today, science is claiming Global Warming and Climate Change to be the medium for the mass deaths of animals in the sky, on the land, and in the waters. Please take note, billions of animals that fly in the sky, those that live on the land and in the sea have already died. As of January 2020, over a billion animals died by the firestorms in Australia alone. And many others in the fires of California and the Amazon of South America.

Third, could it be that the prophetic aspect of Hosea's predictions are being fulfilled in our time? Both the sign of the conditions of people and the mass deaths of animals is so real in the world today. The people of the world appear to have attitudes void of mercy, truth, and a decreasing knowledge of God. But there are more prophecies of the Old Testament. The prophet Joel predicts about 'End Time' events.

Chapter 12

Prophet Joel Predicts Unusual 'End Time' Signs

The prophet Joel received prophetic messages from God which he ministered to the Israeli people. However, his predictions not only had a message during his time, but many of them had prophetic messages for the future. However, some of the prophetic aspects of his message God directed to our time. The messages state that (1) the Messiah will return to the earth, (2) there will be strange and unusual signs in the sky and on the earth, and (3), most of the strange and unusual signs relate to the "Great Battle of that Day of God Almighty at Armageddon. The place which the apostle John called 'Armageddon' the prophet Joel's prophecy identified it as 'The Valley of Jehoshaphat." Let us now investigate.

Joel Predicts Unusual Signs in the Sky and on the Land

One prophecy of Joel that gives information about strange and unusual activities in the sky and on the land is Joel 2:10. It states that **"The earth shall quake before them; the heavens shall tremble: the sun and the moon shall be dark, and the stars shall withdraw their shining"** The prophetic statement has a direct connection to the coming of the Lord as the Messiah, from heaven through the sky to the earth. For example, the sun and moon will salute the coming of the Messiah. Also, His coming back to the earth will cause signs with the sun and moon. The sun will not give its light and the world for a period will be in darkness. Also, within the darkness, the moon will appear as a ball of blood in the sky.

Besides the signs in the sky, there will be a significant earthquake. The prophet Zechariah and the apostle John prophesied about the great earthquake that will take place as Jesus Christ returns from heaven to the earth. Such information is given in the presentation of their prophecies.

The reason we know this prophecy refers to the coming of the Messiah who is Jesus Christ, the Prophet Joel prophesied to the Israeli people that after God removes a northern army and drives them into desolation, the LORD will do great things for the people (Joel 2:20-21). However, take note of other coming blessings the prophet predicted:

1. He will bless the land with rain, pastures and springs in the wilderness, wheat, and oil will overflow, (vs. 22-24).
2. He will restore the years of the past (vs. 25-26).
3. The people will know that He is with them in Israel and that He is the LORD their God (vs. 27).
4. He will then pour out His Spirit on all flesh, their sons and daughters will prophesy, the old men shall dream dreams, and their young men shall see visions (vs. 28).
5. He will show wonders in the heavens and on the earth, with blood, fire, and pillars of smoke (vs. 30).
6. The sun and moon will have signs. The sun will not give any light and the moon will resemble blood (vs. 31).
7. All the above will take place when the Messiah returns to the earth on that "great and terrible day of the LORD's coming (vs. 32). (Joel 2:22-31)

The above seven points show that Joel is speaking about the coming of the Messiah. On such a day, there will be signs in the sky, on the earth, and among the people. Also, it is interesting to know, the prophecy of Joel 2:27- 28 has not been fulfilled. On the day of Pentecost, the apostle Peter referred to Joel 2:28 (Acts 2:15-21). However, the Prophet knew nothing about the Church of Jesus Christ. His knowledge of the outpouring of the Holy Spirit on the Israeli people would be after the Messiah comes to the earth. This section is fully explained in another section of this chapter. Read!

Joel Predicts Unusual Signs Before and After the Battle at Armageddon – in the Valley of Jehoshaphat

Joel Chapter 3:1-21 the prophet foretold the coming judgment of the Gentile nations.

First, the prophet predicts that God will gather all nations to the valley of Jehoshaphat and will judge the nations for their treatment of the

Israeli people. The nations will have to give an account for how they scattered the Israeli people among the nations and parted His land (Joel 3:1-2).

Second, the prophetic text addresses the battle that the apostle John predicts by name, "the Battle of that Great Day of God Almighty, which will occur at a place called 'Armageddon' (Revelation 16:16). In another chapter, it describes John's version of this great battle. However, Joel 2:10–11 explains the signs or events that God gave him about the battle in the Valley of Jehoshaphat. Note:

1. The *earth shall quake* before all people, especially the Gentile Army Coalition. Other scriptures state that this earthquake will be like none that people have ever seen (see Zechariah 14:4-5; Revelation 16:18)
2. The *heavens shall tremble.* This event will involve all the planets.
3. The *sun,* the *moon,* and the *stars* will not give light because the Voice of the Lord will be heard, and His great army will obey and follow his Word
4. The Day of the Lord is great and very terrible, and nothing will be able to stand against Him

The above prophetic text explains the conditions of that Day of the Lord. His arrival to the earth, *first,* there will be a great earthquake that will be greater than any people have ever witnessed. *Second,* the heavens will tremble. The 'tremble' is because The Messiah is on His way to the earth. This is a great day taking place and God wants the entire earth to know it, especially the armies of the Gentile world powers. Third, there will be signs in the sun, moon, and stars. This is more of a celebration that The Messiah is on His way to the earth to defeat Satan, his human armies, restore to the Nation of Israel the promised-blessings, and set up the Kingdom of God on the earth.

Joel Predicts Strange and Unusual Signs with the Heavenly Bodies as the Messiah Returns

Note the five major important points about The Return of The

Messiah to the earth.

First, God will judge Israel and the Gentile nations.

Second, the Nation of Israel will repent and accept their Messiah and will receive the full covenant promised blessings that God had for them.

Third, they will receive the pouring out of the Holy Spirit because they will finally recognize Jesus Christ as their Messiah. The filling of the Spirit of God on them was a promised blessing for them if they accepted Jesus Christ as Lord, Savior, and their Messiah when he first came to them. Remember—he came to his own, and his own did not receive him, but for as many as received him, to them he gave eternal life. Also, he allows them, the believers, to become filled with the Holy Spirit on the Day of Pentecost. On that day, the apostle Peter mentioned this promise that is in the prophet Joel's prophecy—that God will pour out his Spirit on his people. Because the believers of Jesus Christ accepted Him as their Savior, they were filled with the Holy Spirit. Therefore, God partially fulfilled Joel's prophecy.

Fourth, the troubles the Israeli people will endure during the Great Tribulation on the earth, they will be looking for the Messiah. Therefore, as Jesus Christ approaches Mount Olives, the Israeli people will hear His voice from heaven as he returns to the earth. Along with all the signs in the sky and on the land, they will repent of their sins and accept Him as their Savior.

Fifth, therefore, God will fulfill the prophecy in its entirety. Do you think it was a great time on the Day of Pentecost? Well, on this day, Pentecost will be a small matter. Note, as God pours out the Holy Ghost on the Israeli people, (1) The sons and daughters will prophesy. (2) Old men will dream dreams. (3) Young men will see visions. (4) The Servants of the Lord, both men, and women, God will pour out His Spirit. (5) God will show wonders in heaven and on the earth. (6) There will be signs in the heavens and on the earth. Blood, fire, and billows of smoke. The sun and moon will not give light. (7) As a result of all these signs, "everyone who calls on the name of the LORD as He enters the Mount of Olives, will be saved. In addition, the City of Jerusalem will be delivered from the hands and power of the Antichrist/Beast.

Chapter 13

Prophet Amos Predicts Unusual 'End Time' Signs

History shows that Amos was a shepherd before his calling by God to prophesy about future events. After becoming a prophet of God, he was fearless and outspoken. He stirred the people to whom he prophesied. His prophetic ministry appeared to be during the days of Uzziah, king of Judah, and Jeroboam, the son of Joash, king of Israel. We believe that some of his prophetic messages from God were twofold. *First* to the people of Damascus, Gaza, Tyrus, Edom, Ammon, and Moab. *Second,* to the people of Judah. To the second group, he mentioned their transgression and disobedience to God. Also, for such transgressions, God will judge them and the land with darkness, pests, fire, and destruction. The unusual events will occur in the sky, on the land, and the oceans. Yet God will have mercy on the Israeli people. There will be a rebuilding of their dwelling places. Let us now examine these four areas of the prophecies.

Amos Predicts the Signs of Darkness that Will Invade the Sky on the Day of the LORD

Such strange and unusual signs will take place with the Lord's coming to the earth. As the prophet Joel predicts, Amos also saw gloom in his vision, with no brightness at all. Note, "Shall not the **Day of the Lord** be darkness, and not light? even very dark, and no brightness in it?" (Amos 5:20). The prophet left the events related to the day of the Lord in question form. Amos stressed that the 'Day of the Lord' is a dreaded day, and people need to pay much attention to the prophecies about this day, which was the theme of his prophetic message. Bear in mind, the Day of the LORD refers to the day of judgment when the Messiah will come to the earth. Therefore, on that day, or even before, people will see signs occurring in the sky and darkness will cover the earth.

Amos Predicts the Sign of Pests Invading the Land

In the book of Amos, not only do we find activities in the skies, but five main prophecies involve activities on the land.

The Locusts are Coming

Amos predicts the coming of locusts on the land (Amos 7:1-3). God wanted to get the attention of the people of Amos' day. Therefore, he allowed locusts to invade the land. He expected this act to cause the people who had gone away from Him to return. Note, the coming of the Lord has not changed. This event is still a future one. Could it be that the signs of pests attacking the land today also indicate that God is signaling people to come to him because He is coming soon? Could it be that global climate change and its effects are *Signs of the End Time?*

As seen with global climate change, problems are today taking place with pestilence. Famines are on their way as the land continues to heat by the effects of global warming. Jesus Christ predicts that earthquakes, famines, wars, and pests will come upon the land in the *End Time.* These, he taught His disciples, will be some of the signs people must see, understand, and look for His return to the earth.

Today, pests seem out of control. Many mosquitoes and other pests are now in the world. They bring new diseases that science has never seen before. Not only do they transmit viruses, but the scientific community has no cures for these new diseases. Then, some diseases are just showing up and nobody knows where they are originating. For example, the "Coronavirus" is such an example. It is spreading and out of control. These out-of-control diseases have become some of the pestilences Jesus Christ predicts will take place in the *End Time.* Could it be said that people today are now experiencing signs of the *End Times'* events? Let us investigate other prophecies of the prophet Amos.

Amos Predicts a Sign of Fire that Will Have an Effect on the Seas, Oceans, and Land

Note his prophecy about the oceans, seas, and land:

> Thus, hath the Lord God shewed unto me: and behold, the Lord God called to contend by fire, and it devoured the great deep and did cause part of it to disappear. Then said I, O Lord God, cease, I beseech thee: by whom shall Jacob arise? for he is small. The Lord repented for this: This also shall not be, saith the Lord God. (Amos 7:4–6)

Background for God's Prophetic Message to Prophet Amos about a Devouring Fire

The prophet prediction presented is about God's future Judgment of fire that will influence the *seas, oceans*, and the *lands* of the earth. However, before God pronounced His judgment on Israel, the prophet warned them from the very beginning of his messages to them. For example, in chapters one and two, he predicts coming judgments on the nations around Israel. In chapter three he presents witnesses against Israel who condemns the nation for their sins. In chapter four, he condemned Israel for not repenting of their sins and returning to God. In chapter five, he again encourages Israel to repent of their sins they have committed against God. And in chapter six, Amos prophesies against the complacency and pride which Israel portrayed. So, in chapter seven, he predicts (1) Coming Locusts, (2) a Devouring fire, and (3), a Plumb Line. We now discuss *the prophetic coming devouring fire.*

Interpreting the Prophetic Coming Devouring Fire

In observing the three prophecies of chapter seven closely, there is a direct relationship to the continuous sins of the people around Israel, and the people within the Nation of Israel. Therefore, God warns them with signs like the signs seen today with Global Climate Change. So far, this chapter has discussed the 'Pests' or 'Pestilence' the Bible calls it. It now has a discussion on the 'Fire" that will dry up the "great deep" and will devastate the "lands" of the earth. The two-fold message is, (1) to the Israeli people, (2) a prophetic message of some cataclysmic event where a fire will dry up water and destroy the land, and (3), an event that has prophetic inference to the return of the Messiah to the earth.

After Amos cried to God on behalf of Israel, about the nation being so small the judgment did not take place then, in that time of the prophet. However, the prophetic message is, Amos saw in this vision where the 'fire' 'dried up the deep and devoured the land.' The news media has presented many stories where the earth is burning by fires. Science calls them 'wildfires' or 'forest fires' that the warming of Global Warming/Climate Change caused. Again, if the wildfires which are taking place today relate to God's judgment, they will become worse when they dry up the waters and the lands of the earth.

Could it be said that Amos' prophecy about fire affecting the oceans, seas and the lands of the earth is already taking place, and will continue until some greater cataclysmic event takes place? Will fire dry up the "great river Euphrates" when the sixth angel pours out the Sixth Vial of Judgment? Does it all mean that global climate change is an act of God, because of the sins of people and their nations, as seen here in the prophecies of Amos? Yes, it is possible that God is using Global Warming and its Climate Change as a sign of coming judgment on the earth and its people because of sinful practices and denial of Him. God is not willing that any should perish, but that all people should come to repentance and know Him while there is time.

Amos Predicts a 'Plumb Line' Sign – God's Judgment on Israel

The "Plumb Line" shows that God will judge Israel (see Amos 7:17). However, the plumb line also shows 'rebuilding' and hope. Today Israel does not have all the promises that God made with their forefathers. This is because of their rebellion against Him. However, there is the Messianic Kingdom that He must establish. God will fulfill Amos' 'Plumb Line' prophecy in the last- day-set-up of the Messianic kingdom when Jesus Christ returns with his bride and judges the nations of Israel and the Gentiles. Only then, after the Jewish people have accepted Jesus Christ as their Messiah will He set up the kingdom. On that day, they will all turn to him for salvation. Remember— these events of signs will take place at the end of the great tribulation.

Before the setting up of the Messianic Kingdom, Israel will go into captivity. They will lose the joy of their festivals, worship, and famine of not hearing the words of the LORD will take place. The Israeli people will "stagger from sea to sea and wander from north to east, searching for the word of the LORD, but they will not find it. Many other misfortunes will come upon them (see Amos chapters 8:1-14 – 9:1-10).

God told Amos that He will restore Israel (Amos 9:11-15). On that day, He will bring back the Israeli people, His people from exile. The Israeli people will then "rebuild the ruined cities and live in them. They will plant vineyards and drink their wine; they will make gardens and eat their fruit. I will plant Israel in their own land, never again to be uprooted from the land I have given them, says the LORD your God "(Amos 9:14-15).

Chapter 14

Prophet Zechariah Predicts Unusual 'End Time' Signs

First, the events with the greatest *signs ever will be seen in the sky*. The Messiah, Jesus Christ will come through the sky on His way to the earth. During this event, the prophet predicts *signs with the sun in the sky* during the return of the Messiah to the Mount of Olives.

Second, Zechariah predicts a sign that will involve the *weather*. There will be a great earthquake. The earthquake will cause Mount of Olives to split into two parts, forming a great valley leading to the Eastern Gate of the City of Jerusalem. Let us now investigate.

Third, Zechariah predicts that there will be *signs on the land*. For example, (1) a cataclysmic change of restructuring will take place with the Mount of Olives. (2) He will reopen The Eastern Gate that has been cemented for centuries. (3) God will defeat the Gentile nations' armies. (4) Diseases will afflict the Gentile nations.

Fourth, there will be many signs taking place with the *waters on the land.* Streams will flow over the barren lands and crops will be in abundance. However, not only will there be just water flowing, but the unusual event of *living waters flowing* out of Jerusalem over the lands of the earth.

Zechariah Predicts Unusual Signs in the Sky as the Messiah Returns to the Earth

Note the prophecy:

1. Behold, the day of the LORD cometh, and thy spoil shall be divided in the midst of thee.
2. For I will gather all nations against Jerusalem to battle; and the city shall be taken, and the houses rifled, and the women ravished; and half of the city shall go into captivity, and the residue of the people shall not be cut off from the city.
3. Then shall the LORD go forth, and fight against those nations, as when he fought in the day of battle.

4. And his feet shall stand in that day upon the mount of Olives, which *is* before Jerusalem on the east, and the mount of Olives shall cleave in the midst thereof toward the east and toward the west, *and there shall be* a very great valley; and half of the mountain shall remove toward the north, and half of it toward the south.

5. And ye shall flee *to* the valley of the mountains; for the valley of the mountains shall reach unto Azal: yea, ye shall flee, like as ye fled from before the earthquake in the days of Uzziah king of Judah: and the LORD my God shall come, *and* all the saints with thee. (Zechariah 14:1-4)

The Messiah Coming from the Sky to the Mount of Olives, East of the City of Jerusalem

Many of the prophets, Jesus Christ, and the apostles predicted the return of the Messiah, who is Jesus Christ. However, they gave no time or place of His coming. But the prophet Zechariah predicted the exact point of entry of the Messiah. However, to better understand this prophecy, readers need to know some background information about the Messiah, that the prophet Zechariah is predicting about his coming back to earth.

First, we understand that the people had the Messiah taken out of the City of Jerusalem as a criminal after the religious mob condemned to die on a cross at Golgotha. So, it was outside the city of Jerusalem the Roman Soldiers crucified Him. 'Golgotha' in Aramaic means 'place of the skull.' The Latin word for 'skull' is 'calvaria' which today refers to 'Calvary.' Golgotha is an area of the Mount of Olives.

Second, Jesus Christ also ascended from the Mount of Olives. Also, Zechariah predicts that He will return outside the City of Jerusalem, at the Mount of Olives.

Third, when the Messiah returns to the Mount of Olives, the world confederate armies led by the Antichrist will have already conquered, destroyed the people and their properties, and besieged the city of Jerusalem. The Antichrist will also divide the Israeli people. He will send some to other places outside the city and some he will allow remaining in the city. Note Zechariah's prophecy about this day when the Antichrist attacks the City of Jerusalem as seen in the prophecy.

Fourth, note, after The Antichrist will have conquered the City of Jerusalem, he will live there and will control the laws of the land and the people. Remember, according to the apostle John, the Antichrist will want people to worship him as *god.* (see Revelation 13:12&15; 14:9&11)

Fifth, at such time, God will intervene on behalf of Israel. His intervention will be by Jesus Christ the Messiah on His return to the earth and will make war with the Antichrist and his confederacy. As the Messiah is coming through the sky, the entire world will be watching. His coming back to the earth will be with power. The power of His coming will affect the sun, moon, and the stars. Lightning will flash from east to west, says the Bible. The activities in the sky and on land will present the biggest 'storm' the world has ever seen. In addition, the greatest earthquake will take place. Agreeing with Zechariah's prophecy, Revelation 1:7 states, "**Behold, he comes with clouds; and every eye shall see him, and they also which pierced him: and all kindred of the earth shall wail because of him.**" The apostle John has a greater description of the scene of the Messiah coming to the earth. He describes his dress, his possession of saints and armies coming with him. Later in this chapter, John gives this in his prophecies.

Sixth, all the attention by the coming of the Messiah will have the people of the entire world in fear and disarray. However, the Antichrist, the False Prophet, and the World Confederation of Armies that are under the control of Satan will go on the offensive. Satan's controlled armies will await the Messiah's arrival to make war against Him. Revelation 19:19 confirms this act. Note, **"And I saw the beast, and the kings of the earth, and their armies, gathered together to make war against him that sat on the horse, and against his army."**

Seventh, according to Zechariah, the Messiah will touchdown on the Mount of Olives which will get the world's attention. People will probably see the event on television. His very presence in the area will cause catastrophic geological changes. The shaking of the heavenly bodies of the sun, moon, and stars, and the Mount of Olives will cause the formation of a great valley leading to the Eastern Gate of the City of Jerusalem. This is what Zechariah received from God. But in Revelation 16:18-21, it gives greater detail of this event. Not only will the Mountain of Olives split, but all the cities of the earth will fall, every island will flee, and all other mountains, Messiah's entry to the earth will affect. All these strange and unusual signs/events will also cause the same conditions with the weather.

Zechariah Predicts Unusual Signs with the Weather as the Messiah Returns to the Earth

According to the prophet Zechariah, God revealed to him this event of the splitting of the Mount of Olives and the forming of a valley will be a sign. The apostle John, explaining this event of the King of Kings and Lord of lords earth's entry also predicts that there will be strange and unusual signs with the weather. For example, he predicts that there will be an earthquake, along with the many storms of hail, lightning, thunderstorms and so on. Zechariah pinpoints the fact of a particular mountain splitting as the Messiah touches down on it.

The Strange and Unusual Great Earthquake will Split the Mount of Olives and Form a Great Valley

One of the great signs will be with the weather during the return of the Lord Jesus Christ with his bride and the armies of heaven to the earth. Zechariah the prophet predicts the following activities:

> And his feet shall stand in that day upon the Mount of Olives, which is before Jerusalem on the east, and the mount of Olives shall cleave in the midst thereof toward the east and toward the west, and there shall be a very great valley; and half of the mountain shall remove toward the north and half of it toward the south.
> And you will run to the valley of the mountains; for the valley of the mountains shall reach unto Azal: yes, you will run like one running from before the earthquake in the days of Uzziah king of Judah: and the Lord my God shall come, and all the saints with thee. (Zechariah 14:4–5)

When Jesus Christ returns to the earth with his bride, the church (called-out assembly of believers), he will be the Messiah for the Jewish people. Strange and unusual activities or signs will take place with his return. There will be changes in the heavenly bodies. The sun will hold its light, the moon will do the same, and there will be activities with the stars. Also, there will be earthquakes; the Mount of Olives will split into two parts, with a great valley leading to the Eastern Gate. As Jesus Christ predicted, signs will occur with the weather of the sky and earth, as John and Zechariah described (see Revelation 16:16-21). Zechariah not only saw the Mount of Olives splitting and forming a valley, but other weather signs with the waters.

The Sign of Unusual Living Waters Flowing out from Jerusalem Over Desert-Lands

Zechariah saw the flowing of what he called "living waters." He prophesied that the living waters would flow out from Jerusalem. Also, half of these living waters would flow to the *Eastern Sea* and the other half to the *Western Sea*. The flowing of these living waters will continue in the summer and the winter. The presence of the Messiah on earth will change the system of summer and winter, flowing streams, waters, and desert-lands.

Concerning the "flowing of water out of Jerusalem in the summer and winter," Benson's Commentary, Matthew Henry's Concise Commentary, Barnes Notes on the Bible, Matthew Poole's Commentary, Cambridge Bible for Schools and Colleges, Gill's Exposition of the Entire Bible, the Geneva Study Bible, and Pulpit Commentary somewhat agree and state:

> the enlightening, quickening, and saving truths of Christianity, accompanied by the power of the Holy Spirit, shall proceed from the church of Christ, the true spiritual Jerusalem; half of them toward the former sea— The eastern sea; and half of them toward the hinder sea so that nothing shall totally impede its progress, till the Lord shall become King over all the earth.

The author sees the "flowing of waters" from a different perspective. He sees a real flowing of waters coming from the City of Jerusalem to the West, the Mediterranean Sea, and to the East, the Dead Sea. The reason for this interpretation, he states:

(1) The prophecy has nothing to do with Christianity. Jesus Christ will have already taken His Believers. They would have stood at the Bema Seat Judgment, received awards, and taken part in the Praise and Worship of the Lamb around the throne of God. In fact, Zechariah

14:5 makes it clear, **"and the Lord my God shall come, and all the saints with Him."** This *sign* is all about the Second Coming of Jesus Christ to the earth and the blessings that He will give to the Israeli people as He sets up His kingdom. Also, with Him will be His Saints.

(2) The prophecy deals with a total reconstruction of the land area in and around the City of Jerusalem. The presence of the Messiah will change all areas. Note, The Righteous Judge, The Messiah, The King of Kings and Lord of Lords, will be back in the world. Things will not be as we know them. The apostle Paul saw it this way, and he states:

> but just as it is written [in Scripture], "THINGS WHICH THE EYE HAS NOT SEEN AND THE EAR HAS NOT HEARD, AND WHICH HAVE NOT ENTERED THE HEART OF MAN, ALL THAT GOD HAS PREPARED FOR THOSE WHO LOVE HIM [who hold Him in affectionate reverence, who obey Him, and who gratefully recognize the benefits that He has bestowed." (1 Corinthians 2:9)

(3) The changes will be with the "day and night, summer and winter, cold and frost." All have to do with the weather. The weather will be so good that there will be no distinction between day and night. Jesus Christ, the Messiah who is also the "light of the world" will be in the City of Jerusalem. So, there will be no need for the shining of the sun. For sure, the global warming and climate change activities that began all the excitement will end in a blessing for the land of Israel. The behavior of the sun, moon, stars, thunder, lightning, and storms all seem to subside at the Battle at Armageddon. (Revelation 16:16-21)

(4) God cursed the land of Israel because of the disobedience of the people. When Jesus Christ returns the second time to the earth, he will find all the nations fighting against the land of Israel. Jesus Christ will intervene and set them free. The entire nation will turn to him. In response, as predicted by other Old Testament prophets, God will heal the land and its people. Because the Israeli people have now accepted Jesus Christ as their Messiah and King, the Holy Spirit will fall upon them as on the day of Pentecost, thus bringing closure to Joel's prophecy.

(5) Another important thing about the land area, there is a belief that the land where Jerusalem stands may be where the Garden of Eden once was. Also, the sin of Adam and Eve caused God to curse the land

and Paradise was lost. Therefore, the return of the Messiah to the land area, will once again remove the curse from the land. (Read Wellman's article, "Where on Earth was the Garden of Eden?"

https://www.oaoa.com/people/religion/article_7d948064-935d-11e4-90c2- 4b09cc36f260.html)

Having discussed some of the unusual, related signs of the weather, let us now turn our attention to the signs that Zechariah predicts to take place on the land. Note the signs with the weather.

Zechariah Predicts Unusual Signs on the Land as the Messiah Returns

> and the mountain of Olives shall cleave in the midst thereof toward the east and toward the west, and there shall be a very great valley, and half of the mountain shall remove toward the north and half of it toward the south (Zechariah 14:4).

Several major signs or events will occur on the land when the Messiah returns to the earth by Mount of Olives. Of these many events, discussed here are (1) the Opening of the Eastern Gate of the City of Jerusalem, and (2) the Unusual Event of a Strange Disease that will Plague the Soldiers who Fought Against Jerusalem.

The Unusual Event of the Opening of the Eastern Gate to the City of Jerusalem

The cataclysmic change of reconstruction of the area will have a great effect on the opening of the Eastern Gate to the City of Jerusalem. The geographical shifting patterns of this mountain means that the newly-formed valley lies from east to west. Therefore, the Messiah, Jesus Christ, His armies, and Saints will enter the City of Jerusalem through the Eastern Gate. We must note here, the Eastern Gate is closed and will remain closed until the Messiah returns. It means that during the return of the Messiah to the Mount of Olives, He will only have to speak to the Eastern Gate, and it will open.

God Closed the Eastern Gate after His Glory Departed from Jerusalem

First, although Zechariah did not share this information, the prophet Ezekiel did. He predicts that, in a vision, God brought him to the Gate which led to the sanctuary which looks to the east; but God had shut the gate. God explained to him that the gate will not open until the Prince enters, sits (on the Throne of David), and will eat bread before the Lord. (Ezekiel 44:1-3).

Second, remember that before the time of the prophet Zechariah, God gave a prophecy to the prophet Ezekiel that there would be the closure of the Eastern Gate and would not reopen until the Messiah entered at the gate. Gillette (2017) states that "approximately 2,600 years ago, the prophet Ezekiel received a vision of the Glory of the Lord" that had to do with this event.

Third, prior to this prophecy, Ezekiel God's glory and the actions of the cherubim. He saw the glory of God leaving **"the door of the east gate of the Lord's house, and the glory of the God of Israel was over them above."** (Ezekiel 10:19)

Fourth, these prophecies show the importance of the Lord God Almighty and the importance of this Eastern Gate at the city of Jerusalem. God has a connection here. Here is a little history of the partial fulfillment of this prophecy.

The Eastern Gate faces the Mount of Olives. It is one of eight gates to the city of Jerusalem and the only gate that allows traffic at an entrance from the east. Per this prophecy, the Eastern Gate of Jerusalem is to remain closed until the Messiah enters.

In 1517, when the Turks conquered Jerusalem under the leadership of Suleiman the Magnificent, he commanded the rebuilding of the city's ancient walls. During this rebuilding project, for some unknown reason, he ordered the sealing up of the Eastern Gate with stones. In addition, he put a Muslim cemetery in front of the gate, believing that no Jewish holy man would defile himself by walking through a Muslim cemetery, trying to enter the gate (Reagan 2014).

God will open the Eastern Gate for the Messiah to Bring Back His Glory to the City of Jerusalem

First, prior to the closing of the Eastern Gate, Jesus Christ fulfilled one prophecy of the prophet Zechariah. It states, **"Rejoice greatly, O daughter of Zion; shout O daughter of Jerusalem: behold, thy King cometh unto thee: he is just, and having salvation; lowly, and riding upon an ass, and upon a colt the foal of an ass"** (Zechariah 9:9). John 12:12–16 shows the scene of Jesus Christ's triumphant entry into the City of Jerusalem. This act of Jesus Christ clarifies that one-day God will have the Eastern Gate open again and the people will rejoice as they did with Jesus' triumphant entry then.

Second, the triumphant entry of Jesus of Nazareth fulfilled a prophecy that Zechariah predicted over five hundred years earlier (Gillette 2017). The unusual event of Jesus Christ's triumphant entry into the city of Jerusalem through the Eastern Gate was a partial fulfillment of the prophecy. He is coming again, and this time, as the King of kings and Lord of lords, per the prophecy of Zechariah. He will not be riding on a colt, but on a white horse, all the way from heaven to the Mount of Olives, east of the City of Jerusalem. Although the Muslims under Suleiman the Magnificent cemented the gate closed, it will open on that day as the Messiah enters the earth.

Third, the Messiah Jesus Christ's entry into the City of Jerusalem as King of Kings will meet much hostility. The Beast or Antichrist, his False Prophet, and the armies of the Gentile world will meet him for war. However, during this day of the battle of God Almighty, at Armageddon, the prophet Zechariah saw an unusual event taking place. A strange disease plagued the soldiers who were fighting against Jerusalem.

The Unusual Event of a Strange Disease that will Plague the Soldiers Who Fought Against Jerusalem

The prophet Zechariah predicts that strange and unusual diseases will plague the land. A terrible disease will afflict people and animals. (Zechariah 14:12–14)

Remember such a plague of this type has not entered the world yet. This onset of the disease is a futuristic, strange, and unusual form of sickness that the prophet is predicting will come on the animals and people of the earth. For example, with this scourge, the flesh of the people will rot while they are actively standing. They will not have a long-standing illness. In a moment, while standing or walking, their flesh will rot and fall off the body. Also, their eyes will fall out of the eye sockets. The plague will also affect animals, such as horses, mules, donkeys, camels, and all other animals.

Interpreting this prophecy is a prediction that sometime soon a confederation of nations will come and fight against Jerusalem and the people of the nation of Israel. This event has not yet taken place but will happen. After the people of Israel receive this treatment from the nations of the Gentiles, this unusual and strange disease, which the prophet is predicting, will come to the animals and people of the earth. "Camels" seem to show the location of this problem of sickness. Few people ride and use camels in Western countries.

The nation of Israel has not been a favorite to many other countries in the world. Many of Israel's enemies would not recognize her as a nation. However, she will become a sign within herself to the nations around her. Some signs are already taking place. For example, not long-ago Russia moved into Syria, which is near the land of Israel. Predictions suggest that a country of the north will one day come and lay siege against Jerusalem. However, whatever the case may be, the Bible predicts that God will punish the people or nations that will declare war on Jerusalem with this strange and unusual plague of sickness.

Remember, this chapter has presented and explained the Old Testament predictions or prophecies regarding strange and unusual signs that will take place in the End Time. However, the Old Testament prophets were not the only group of predictors in the Bible. In the New Testament, there are many prophecies in the GOSPELS made by the Lord Jesus Christ about strange and unusual events or signs that will take place in the "end of the world." Also, in the New Testament's EPISTLES, the apostles Peter, Paul, and John also predicted End Times' Signs. You must not miss these. Read Sections Four and Five!

SECTION FOUR

THE NEW TESTAMENT GOSPELS PREDICT UNUSUAL 'END TIME' SIGNS

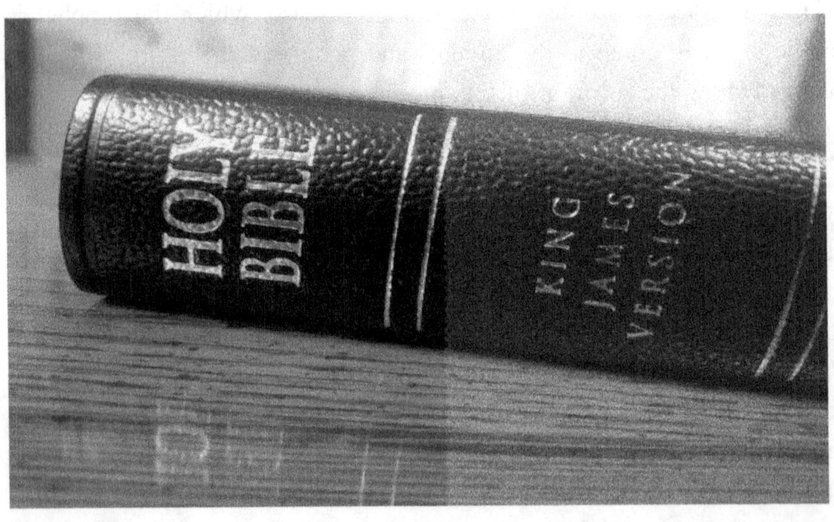

~✦~

The prophecies of 'End Time' events are the predictions of Jesus Christ that He called "Signs." The Gospels of the New Testament of the Bible – Matthew, Mark, and Luke, are the Gospels books used to present Jesus Christ's End Times' predictions. Before we present *The Prophecies of Jesus Christ*, it is good that readers understand some of the *background* and the *work* of The Person–Jesus Christ, and His apostles whom He called into the Gospel Ministry, who recorded His work. Therefore, Section Four presents:

First, a background on (1) The Person-Jesus Christ; (2) The Work of Jesus Christ on Earth and His Predictions; and (3), The Apostles of Jesus Christ and their Recordings of His "End Time" Signs.

Second, it presents the predictions of Jesus Christ as recorded by apostles Matthew, Mark, and Luke. Let us now read about the background, then the three section-chapters of Matthew, Mark, and Luke's writings of Jesus' predictions of End Times' Signs.

The Person–Jesus Christ

The Book of Hebrews, chapter one, God introduces His Son, Jesus Christ. The same One who is Builder and upholder of the universe, Savior, and the coming King of Kings and Lord of lords, and Messiah. Note the introduction God gives:

> God, who at sundry times and in divers manners spake in time past unto the fathers by the prophets,
> He has in these last days spoken unto us by his Son, whom he hath appointed heir of all things, by whom also he made the worlds.
> Who being the brightness of his glory, and the express image of his person, and upholding all things by the word of his power, when he had by himself purged our sins, sat down on the right hand of the Majesty on high:
> Being made so much better than the angels, as he hath by inheritance obtained a more excellent name than they.
> For unto which of the angels said he at any time, Thou art my Son, this day have I begotten thee? And again, I will be to him a Father, and he shall be to me a Son?
> And again, when he bringeth in the first-begotten into the world, he saith, And let all the angels of God worship him.
> And of the angels, he saith, Who maketh his angels, spirits, and his ministers, a flame of fire.
> But unto the Son, he saith, Thy throne, O God, is forever and ever: a scepter of righteousness is the scepter of thy kingdom.
> Thou hast loved righteousness, and hated iniquity; therefore God, even thy God, has anointed thee with the oil of gladness above thy fellows.
> And Thou, Lord, in the beginning, hast laid the foundation of the earth; and the heavens are the works of thine hands:
> They shall perish, but thou remainest, and they all shall wax old as doth a garment.
> And as a vesture shalt thou fold them up, and they shall be changed: but thou art the same, and thy years shall not fail.
> But to which of the angels said he at any time, Sit on my right hand, until I make thine enemies thy footstool?

Are they not all ministering spirits sent forth to minister for those who shall be heirs of salvation?

First, the above chapter reveals that years ago, God communicated with people through the ministry of the prophets. God did so with different methods and at different times.

Second, in these *End Times* God speaks through His Son, Jesus Christ. People need to understand that according to this chapter, The Triune God made an agreement in heaven. In the Hebrew language, the language of the Old Testament, the word 'God' is 'Elohim' which is a plural noun that means "three or more." Therefore, it is evident that in the Godhead, the Father is God, the Son is God, and the Holy Spirit is God. Therefore, One of His Persons will be God the Father, another Person will be God the Son, and the other Person will be God the Holy Spirit. The Godhead did this way, so that people may be able to understand God regarding their Salvation and His Saving Grace.

Third, God appointed Jesus Christ heir of all things. In fact, God made the heavens, earth, and all other planets through God the Son, Jesus Christ. See, when any person sees Jesus Christ, he or she is looking at the very image of God (see John 14). Jesus Christ is the brightness of the Glory of God. Jesus Christ is the person who upholds all things. He does it by the power of the words He speaks. Jesus Christ is the One who has purged our sins. After he completed this plan of Salvation, He sat down on the throne at the right hand of God, the Majesty in heaven.

Fourth, to the Son, Jesus Christ, note what God said, "Thy throne, O God, is forever and ever: a sceptre of righteousness is the sceptre of your kingdom. You have loved righteousness, and hated iniquity; therefore I, God the Father, even thy God, I have anointed you with the oil of gladness above all of my creation. And You, My Lord, in the beginning, You laid the foundation of the earth. The heavens are the works of your hand. One day the heavens will perish, but You are forever. Also, the things you made will one day perish, but you will never change. You are the same and will always remain this way.

Fifth, the above scriptures speak solely to the coming of Jesus Christ into the world to die for the sins of humankind. On his arrival into the world, all the angels of God worshiped him. God is the only one to worship. Yes, Jesus Christ came to earth. He was born of the Virgin

Mary, grew up, became a man, and at twelve years old began his ministry in the land. During his life on earth, he spoke of and taught people about God, and he prophesied about future events that will be signs of the end time. Often, men would approach him with questions. Like most of us, people seem to have always sought a sign that offers some understanding of life or one they may follow. In fact, this was a common practice during times past, and even today we look for signs, especially those that show what will happen.

The Work of Jesus Christ on Earth and His Predictions of 'End Time' Signs

First, the work of Jesus Christ on earth was to make peace by destroying the barrier that was the dividing wall of hostility between God and people. Jesus Christ accomplished the task by His death, resurrection, and removing the commands and regulations of the law. He preached peace to those who were far away and those who were near. By preaching this peace, he bridged the gap between grace and the law and gave all people the opportunity to come to the Father by the one Spirit, the Holy Ghost.

Second, He made it possible for people to no longer be foreigners and strangers but to become citizens with the household of all God's people. According to the Bible, the household is "built upon the foundation of the apostles and prophets, Jesus Christ himself being the chief cornerstone" (Ephesians 2:20).

Third, during Jesus Christ's ministry on the earth, He made many predictions of the 'Last Days' or the 'End Times'. The recordings of the first three books of the New Testament, in the Bible, are the prophetic works of Matthew, Mark, and Luke. Because the first three books are similar in the presentation of Jesus Christ's work on earth, they are the Synoptic Gospels. Therefore, in Section Four, the presentation of the prophecies are from the Synoptic Gospels - Matthew, Mark, and Luke. Although there is a Book within the Gospels, the Book of John, all prophecies of the apostle John along with Peter and Paul are presented in Section Five–the Epistles Predictions.

The Apostles Of Jesus Christ and their Recordings of His 'End

Time' Signs

First, according to the Bible, God's Holy Word, the Holy Spirit, when He came, gave gifts to men. Among those that received gifts from the Holy Spirit were the eleven apostles, and the apostle Paul whom Jesus Christ selected, is the twelfth apostle.

Second, the Book of the Acts of the Apostles is a recording of their works for Jesus Christ, thus, building His Church on the earth. Their work was mainly to share the good news of the Gospel of Jesus Christ so that people could come to know Jesus Christ as their Lord and Savior. Jesus Christ, before he ascended to heaven, he gave them this commission to go into all the world and to teach and preach His Message.

Third, some of the apostles of Jesus Christ received *prophetic messages* through the ministry of the Holy Spirit. Today the Bible shares their writings which came from God, through the Holy Ghost who gave them the authority to write and share His Word. We find evidence of this revelation in 2 Peter 1:21. It states, **"For the prophecy came not in old time by the will of man: but holy men of God spoke as the Holy Ghost moved them."** Interpreting this scriptural text taken from the Holy Bible lays down certain underpinnings. Note:

- Prophecy has nothing to do with the human mind or human thinking.
- The prophets and apostles who were humans, The Holy Spirit, The third Person of the Godhead directed them.
- Being in the Spirit of God, the prophets could receive the prophecy from God that he wanted people to hear and take note.
- Bible prophecy ceased when the written Word was completed. Note, *for we know in part, and we prophesy in part. But when that which is perfect comes, then that which is in part shall be done away* (1 Corinthians 13:9-10). The Bible today has all the prophecies of the End Time Signs.

Some of these men to whom God spoke prophetic truths were followers of Jesus Christ. Later, they became apostles. The Holy Bible states why God chose them as apostles. *First,* they must have worked with Jesus Christ in His earthly ministry. *Second,* they had to be eyewitnesses to his resurrection from the grave. These, for example, watched him as he ascended into the clouds from the earth. *Third,* prior to his ascension, he commissioned them for their ministry to go into all the world and preach the Gospel. This is the criteria for becoming an APOSTLE.

Judas Iscariot rejected Jesus Christ and turned his back on his role as a disciple. The early church of the followers of Jesus Christ elected Mathias to fill the role of Judas Iscariot. However, Jesus Christ selected the apostle Paul on the Damascus Road. Paul, in his Damascus-road-experience, was *first,* an eye-witness of the resurrected Jesus Christ. *Second,* in his personal experience, he communicated with Jesus Christ that day. *Third,* he was selected by Jesus Christ to go and minister to the Gentiles. In 1 Timothy 1:1, Paul gives his testimony. Note: **"And I thank Christ Jesus, our Lord, who hath enabled me, for that he counted me faithful, putting me into the ministry."** Again, in Ephesians 2, Paul established (1) his reliance on Jesus Christ, (2) the purpose of His work on earth, and (3), the power of Christ's work in the hereafter (read Ephesians 2:14–20).

Jesus Christ and His apostles are the main persons of the Gospels and the Epistles of the Bible. Matthew, Mark, and Luke predicted 'End Time' prophetic signs in the Gospels. Peter, Paul, and John predicted 'End Time' signs in the Epistles. These two sections of the book, *The Gospels* and *The Epistles* give prophetic predictions of End Time Signs which will occur in the sky, on the land, in the waters, and with the weather. However, Section Four now presents the End Time predictions of Jesus Christ from the Gospels of Matthew, Mark, and Luke. Matthew's Gospel now presents Jesus Christ's predictions of End Time Signs.

~⸸~

Chapter 15

Matthew's Gospel Presents Jesus Christ's Predictions of Unusual 'End Time' Signs

Introduction:

The book of Matthew tells the story and events of the present times and the *End Times*' signs that relate to the Israeli people. However, the predictions made here by Jesus Christ are of great concern to the nation of Israel, but also all people and nations of the earth. With this foundation in mind, before these predicted events take place, many years will expire since Jesus' work on earth.

First, there will be *the church age,* which concludes with the '*Rapture*' or '*Snatching Away*' *of the called-out assembly of believers*.

Second, there will be *seven years of great tribulation*, in which many strange and unusual signs will take place in the sky, on land, and in the waters. Also, many events will take place in the weather that will be unusual signs.

Third, many signs will take place before Jesus Christ returns as the Messiah and King of kings and Lord of lords.

Fourth, many other unusual signs will take place as the Messiah approaches the earth that will affect the sky, the land, the waters, and the weather.

Fifth, Jesus Christ predicts to his disciples these signs of the *End Times* in Matthew 24:1-51. Let us now investigate the *End Times*' signs of strange and unusual events that Jesus Christ predicted as He sat with His disciples on the Mount of Olives, Jerusalem.

Events that Led to Jesus' Disciples' Questions about the *End*

Times' Signs

First, one day as Jesus Christ was teaching, the philosophers of that time, the Pharisees, and the Sadducees came to him, inquiring about *signs*. Here is the story that the apostle Matthew presents in the book of Matthew:

> The Pharisees also with the Sadducees came, and tempting, desired him that he would shew them a sign from heaven. He answered and said unto them, When it is evening, ye say, It will be fair weather: for the sky is red. And in the morning, It will be foul weather today: for the sky is red and lowering. O ye hypocrites, ye can discern the face of the sky; but can ye not discern the signs of the times? A wicked and adulterous generation seeketh after a sign; and there shall no sign be given unto it, but the sign of the prophet Jonas. And he left them and departed. (Matthew 16:1–4)

Jesus Christ's answer to this group of philosophers showed that they had the ability to discern the *signs of the skies* but could not understand the *signs of the times* in which they lived. He further told them that the *only sign* they would get from him was *the sign of Jonah the prophet.*

Second, the story of the conversation between Jesus Christ and the scribes and Pharisees did not leave the minds of his disciples. As the disciples and Jesus Christ were walking together, they had a conversation about the beauty and structure of the buildings in Jerusalem. In response to their conversation, Jesus told them that the buildings they were talking about, He predicted their destruction.

Third, when they reached the Mount of Olives and had sat down, his disciples had a personal talk with Jesus about the destruction of Jerusalem he'd predicted. Look at what they said: **"Tell us, when shall these things be? and what shall be the sign of thy coming, and of the end of the world?"** (Matthew 24:3b). Jesus's disciples had a threefold question: (1) *when will the destruction come to the buildings;* (2) *what would be the signs of Jesus's coming;* and (3) *what would be the signs of the end of the age/world?*

Fourth, Jesus Christ's answers to these questions allowed him to present various predictions regarding the *end of the age* and his return to the earth the second time. Note, the discussion is not about His Return in the clouds during the *Rapture of the called-out assembly of believers*. The *End Times' signs* discussed in the chapter are about the 'Return of Jesus Christ' to the earth the second time to judge and set up His Earthly Kingdom.

Jesus Christ Predicts Unusual Signs to Take Place in the Sky

The signs that Jesus Christ prophesied to take place in the sky has to do with the return of Jesus Christ to the earth. He will come back as, (1) King of Kings and Lord of Lords. (2) He will come back to the earth as the Messiah to the Israeli people. (3) He will come back to earth to judge the nations. (4) He will come back to the earth to set up His Kingdom on the earth and will rule and reign for 1000 years. Let us now investigate his predictions about the unusual events that will take place before as He returns to the earth. Matthew 24:29–31 states:

> Immediately after the tribulation of those days shall the sun be darkened, and the moon shall not give her light, and the stars will fall from heaven, and the powers of the heavens shall be shaken: And then shall appear the sign of the Son of man in heaven: and then shall all the tribes of the earth mourn, and they shall see the Son of man coming in the clouds of heaven with power and great glory. And he shall send his angels with a great sound of a trumpet, and they shall gather together his elect from the four winds, from one end of heaven to the other. (Matthew 24:29–31)

*Explaining **the Sign** of Jesus Christ Coming Back in Clouds through the Sky*

This prophecy is about the literal return of Jesus Christ as the Messiah to the earth. This return of Jesus Christ is *the Second Coming of Jesus Christ*. The *Second Coming* involves two distinct parts. The *first* part has to do with his return to the clouds of the skies to receive the called-out assembly of believers. The believers are the born-again followers of Jesus Christ who look for his imminent return. In biblical teaching, this event is the <u>Rapture</u> or the *'Snatching Away'* of the believers. During this event, there will also be many strange and unusual events taking place. In the author's book, "THE MYSTERY: Humans Becoming Immortals," it explains fully this event which will occur with many signs and wonders. The book is available on Amazon, under "Books by Desmond M Coverley, Ph.D. However, in staying with Matthew's account of Jesus' predictions we are on the second section of the 'Coming' of Jesus Christ to the earth.

The second part of the Second Coming is the actual return of Jesus Christ to the earth. This time he will come through the sky with his bride, the believers that he took away in the first part of the Second Coming. Following him and his bride will be his heavenly army. He comes this time to judge the nations, as the Messiah to make things right for the Jewish nation, and as the King of kings and Lord of lords to set up His millennial kingdom.

*The **Unusual Signs** as Jesus Christ Comes through the Sky on His Return to the Earth*

Matthew 24:31 states, **"And he shall send his angels with a great sound of a trumpet, and they shall gather together his elect from the four winds, from one end of heaven to the other."**

When Jesus Christ comes for His believers, they will meet Him in the clouds and not to the earth. However, there are similarities in both phases of His Coming.

First, in His coming for His Believers, (1) the Trump of God will sound. (2) The Archangel will blow the trumpet. (3) The sound will awaken the Believers who died. They along with the living Believers, Jesus Christ will transform them and immediately the cloud will take them to meet the Lord in the air.

Second, the coming of Jesus Christ, the Son of Man to the earth, according to the above scripture in Matthew, (1) a loud trumpet will sound by more than one angel. (2) the sound will gather *His elect* from the four winds, from one end of the heavens to the other." This scripture is not about the Believers of the church of Jesus Christ. It is about "the children of thy people" God told Daniel the prophet. The **"many of them that sleep in earth's dust shall awake, some to everlasting life, and some to shame and everlasting contempt**" (Daniel 12:1-2).

Third, "the shame" which Daniel saw, Mathew describes as the people of the earth mourning when they see the Son of Man coming on the clouds of heaven with power and great glory.

Fourth, His coming to the earth incorporates fame, power, and glory. The peoples of the earth will mourn when they see the Son of Man coming on the clouds of heaven, with power and great glory. Besides lightning-like signs, there will be angels blowing trumpets, which will be thunderous. The prophecy states, **"And he shall send his angels with a great sound of a trumpet, and they shall gather together his elect from the four winds, from one end of heaven to the other"** (Matthew 24:31).

Fifth, this event will not be a small thing. It is a great event. Can you imagine seeing the Son of Man, also known as the King of Kings and Lord of Lords coming from the sky riding on a white horse? Also, with Him thousands and thousands of His Saints who He had already raptured when He met them in the air, just before the Great Tribulation on earth began? Also, with Him, His heavenly armies. Note, the Believers and all the heavenly hosts riding through the sky on horses coming toward the earth. As we read the prophecies of the apostle John, he gives greater description to this Second Coming of Jesus Christ to the earth.

Sixth, additional signs taking place will be, (1) signs will occur in the heavenly bodies. The sun will become darkened, and the moon will not give its light; the stars will fall from the sky. Besides significant planetary activity, there will be thunderstorms and lightning. The illustration Jesus gave the disciples referenced lightning flashing with speed from the east to the west. As lightning that comes from the east is visible even in the west, so will be the coming of the Son of Man.

The ***Sign of the Unusual Time*** *of the Return of Jesus Christ from the Sky*

No one knows the year, month, week, day, hour, a minute or second when Jesus will come back, but there is a clue to consider 'His Return'; He gave many signs. Scripture doesn't tell us the time when He will come, but there is a positioning of the event. Matthew 24:29 says it will be **"immediately after the tribulation of those days,"** meaning the days of the great tribulation on earth, and there will be strange and unusual signs in the sky that **"all the tribes of the earth mourn, and they shall see the Son of man coming in the clouds of heaven with power and great glory"** (Matthew 24:30b).

Before he comes with great glory and power, Jesus Christ confirms that there will be activities in the heavenly bodies accompanying his return to the earth. Such signs will be unusual and strange. There will be a darkening of the sun, a failure of the moon to give light, stars falling from the sky, and the shaking of the heavenly bodies. These activities will be like the sign of the fig tree that foretells that summer is near. These events or signs of earthquakes, famines, sickness, and disease are signs similar to the fig tree sign. The strange and unusual signs of climate change today may mean that Jesus Christ's return to the earth is near. So, if His return as the Messiah of the Jewish people is near, then His return for the believers is closer than people know.

Jesus Christ Predicts Unusual Signs to Take Place on Earth

Jesus' prophecies stated that before He returns to the earth, *First,* people will see unusual signs among themselves throughout the world. *Second,* unusual signs of *Deception* will be clear in the *End Times*. *Third,* there will be unusual events of hate and wars among the Nations of the World. *Fourth*, unusual events of *Disease and Pain* will come upon the world. *Fifth,* there will be unusual signs occurring on the earth during the Great Tribulation. *Sixth*, unusual events will affect the Israeli people. *Seventh*, Jesus predicts the unusual interpretation of the sign of the Fig Tree as a sign of His Coming back to the earth. Let us investigate Jesus' predictions of these *"End Time Signs."*

Unusual Signs of Disbelief among People throughout the World

Jesus Christ predicts that the spiritual condition of people before He comes back to the earth will be at an all-time low. People will move away from God and become self-centered. Jesus Christ likened people's conditions in the *End Times* to be like those days of Noah. Matthew 28:37–39 predicts, **"But as the days of Noah were, so shall also the coming of the Son of man be...eating and drinking, marrying, and**

giving in marriage…so shall also the coming of the Son of man be."

There is nothing wrong with eating, drinking, and getting married. Rather than praising God and acknowledging him, people will live lives that leave God completely out. Thus, when God intervenes, many will not be ready and will perish, as happened in Noah's day. However, the events that people will witness in the sky, on the land, in the waters, and with the weather will be signs. So, to ignore the warning signs of God that danger is coming and live an unconcerned life results in poor outcomes. Noah preached for 120 years, warning humankind of the coming flood. He even built an ark for all to see, but they paid no attention to Noah. However, the people perished in the flood. Jesus likened these days of Noah to days, which are coming to the earth. He also said that humankind will make a similar mistake.

Unusual Signs of Deception Is Clear in the End Time

Jesus Christ says deception is a common trick that humankind will use in the *End Times* or just before he comes back to earth. He warned his disciples to watch out for deceivers who will come. **"And Jesus answered and said unto them, Take heed that no man deceives you. For many shall come in my name, saying, I am Christ; and shall deceive many"** (Matthew 24:4–5).

First, the emphasis on this deception that Jesus Christ warned about refers to his believers. He was explaining to them how his followers will have problems with deception. The fraud will be prevalent among the religious leadership. Before he comes back to this earth, the road of Christianity will have many deceivers. Some will even say that they are the Messiah. In this warning, there is another message, this one to the Jewish people. They have never been certain as to whom their Messiah is or will be. Jesus Christ came, announcing that he was the one, but they rejected him. He warned his disciples about the deceivers that will come, claiming to be the Jewish Messiah. This deception about the Messiah will be a trick on many of the Jewish people. All believers, however, whether born-again individuals who are following Jesus Christ or Jewish individuals who are looking for the Messiah to set them free, should know that false persons will come in the name of God, claiming to be the Christ, and they will deceive many who believe.

Second, Jesus predicted that people would show their deception by the lack of love among themselves. This sign will be prominent among religious leaders. Note His prediction, **"Then shall they (religious people) deliver you (the believers) to the authorities. They will beat, persecute, and even kill you. All nations will hate you because of my name. And the hate will offend many of the followers. So, they will betray one another and will hate one another"** (Matthew 24:9–10).

This prophecy had a lasting effect on the believers of Jesus Christ. The hatred by nations of the world targeted His disciples. Since Jesus predicted this prophecy, many of His followers were hatred and tortured. Yet many who the authorities will persecute will take a stand for Christ. In fact, the Holy Spirit recorded such a historical record of the believers in Hebrews 11, that explained what some of His Believers went through. Note:

> And others had trial of cruel mockings and scourgings, yea, moreover of bonds and imprisonment: They were stoned, they were sawn asunder, were tempted, were slain with the sword: they wandered about in sheepskins and goatskins; being destitute, afflicted, tormented; (Of whom the world was not worthy:) they wandered in deserts, and in the mountains, and in dens and caves of the earth. (Hebrews 11:32–38)

Third, the above was about *deception* and *persecution* then. However, persecution continues on the believers of Jesus Christ throughout the world. As the coming of Christ draws nearer the Believers must know of such hatred, for, Satan wants to kill and destroy all. Many believers will continue to go through persecution and will experience hatred from the world. But they must stand and stand fast. The Book of Hebrews 11:39 states that many believers of the past went through torture and have not yet received the promise God has for them. They died without the promise of God because His program has not yet concluded. The former believers cannot receive the promise of God without the believers of today. Therefore, believers today need to, "…run with patience the race that is set before us. Looking unto Jesus the author and finisher of our faith…Lift up the hands which hang down, and the feeble knees; And make straight paths for your feet…Follow peace with all men, and holiness, without which no man shall see the Lord: Looking diligently lest any man fails of the grace of God; lest any root of bitterness springing up trouble you, and thereby many be defiled." (read Hebrews Chapter 12)

Even though these comforting words are in the Holy Bible, Jesus Christ predicted that among the believers—or rather, those who call themselves followers—the national hatred toward them will turn some into other believers. Some will even become weak in their faith, and many will turn away from the faith and betray and hate each other, as predicted by Jesus Christ in Matthew 24:10. Remember—these predictions by Jesus Christ describe the reality of the *End Times* for believers. Those who follow Jesus Christ will undergo persecution. They will suffer much hatred and betrayal from the so-called believers.

Fourth, besides the betrayal of the Believers by the people of the world, Jesus Christ predicted that some believers would betray other believers because of their lack of love for others. As a result, some will lose their stand for Jesus Christ. But in the *End Times*, many believers will stand firm, and the gospel will continue to reach the entire world as a testament to all nations. Jesus clarified that acts of proclaiming the Word of God will go throughout the world. After such proclamation, "then, the end will come,' stated Jesus Christ.

Unusual Signs of Wars and Rumors of Wars Among the Nations of the World

Jesus Christ predicts that a significant series of "**wars and rumors of war-events. Nations will rise to war against nations and kingdoms against kingdoms"** (Matthew 24:6–8).

National and World Wars

Since Jesus Christ's prediction about wars and rumors of wars, there have been several wars and conflicts in the world. Some countries are at war with each other as we read. In fact, there have been two great world wars through which thousands have died. There were and are continuing conflicts and wars where many others have died and are being killed as you read. Regardless of whether your country has a king, there will be wars because of the human heart and the leadership that governs nations. Wars will take place among the nations of the world. The news media presents such events almost daily. Also, the closer the day gets to the coming of Jesus Christ's return to the earth, the news media will present information on wars and rumors of wars throughout the world. Note, before a war starts, there are rumors or talks about wars. Then, even while a war is going on and after, there are still talks of wars among people. People tell stories of missiles and bombs coming from the skies and from submarines and warships, and how they escape.

So was the case with Mr. Putin of Russia. There were rumors and news about the war he was threatening against Ukraine as he had his soldiers and equipment lined up outside the borders of the nation of Ukraine. Then the time came, and he attacked the country. As a result, millions of people became refugees. Cities and towns are being destroyed as of March 25, 2022. Thousands of soldiers have been killed since one month in the war. Also, it was noted that there was no reason for this devastation. Many of the nations' leaders stated that Europe has not had such a war like this one, since World War II.

Jesus Christ predicted signs of "wars and rumors of wars, nation against nation, and kingdom against kingdom." Since Jesus' war prediction, many forms of war have become known to humankind. For example, adding to the world wars and national conflicts, there are (1) Religious Wars, (2) Drug Wars, and (3) Cyber Warfare.

Cyber Warfare

Besides the old-time wars, a new form of war is being conducted with the use of computer technology—*cyber warfare*. Cyber warfare involves governments of various countries. Each is trying to steal the other's secrets of commerce, trade, and security. Some take part in fighting other countries' electoral processes, as seen in the recent 2016 US presidential election. There are many stories told on TV, in the daily papers, and other means of news about cyber warfare. Apparently, all nations are engaged in this form of warfare. Each wants to learn about the other and how a nation can bring down another through cyberwar.

Religious Wars

There are *religious wars*, one religion against the other. This kind of war spreads much hatred among humankind. People are becoming suspicious of their neighbors because of different skin colors or ethnicity. Thus, fear is everywhere among people, even in places of worship. Note, these problems are occurring within "Churches." Take for example this true story.

In a church which is not named, the congregation was mainly of one color with only a few families of another color. A colored woman was fully involved in the congregational activities. However, on one occasion, she sat next to two ladies of the majority color. After the service, the ladies of the majority color left. However, on leaving, by mistake, they took each other's bags. One of the ladies realized she did not have her particular bag, went to the pastor of the church, and accused the lady of color of stealing her bag. The pastor called the woman and accused her of stealing. Later, it was revealed that both ladies of the majority color had mistakenly taken each other's bag. This plot which Satan planted within this congregation caused the few families of color to leave the church, never to return.

The case above shows how Satan is working even within the churches to prevent the Gospel of Jesus Christ from reaching all people regardless of color, race, or creed. We must not forget the teachings of the Bible that God is interested in all nations, kindred, tongue, and people. In Matthew 24:14. Luke 24:47, and Galatians 3:8, the Bible states that the gospel must go to all nations. Then, in Revelation 5:9; 7:9; 10:11; 11:9; and 14:6, all passages of scriptures state the gospel being preached to "every nation, and kindred, and tongue, and people."

Why does the Bible state multiple times that there is no difference with people? *First,* God wants everyone to be saved. *Second,* God knows that there is a coming Antichrist who will make war with the tribulation saints. *Third,* He knows that there is a <u>Big Religious System which the Bible labels</u> <u>as Babylon the Great Harlot</u> (Revelation 13:7; 17:15). This system has for centuries killed and destroyed many of the people of God. Today it has crept into the church of Jesus Christ, corrupting it, and declaring war on the truth of Jesus Christ. Many believers have been deceived and are still being deceived by this system. However, in Revelation chapters 16-19, it describes how God will destroy this Religious System of Babylon and those who follow such a system. Believers, Beware, cease from warring with your brothers and sisters of other creeds, races, and colors. Christ is coming soon for all who believe in Him. "We shall be all changed, in a moment in the twinkle of an eye." (1 Corinthians 15:52).

Drug Wars

There are *drug wars* over territory and the sale of narcotics. Within these wars are many other factions of wars. Note, the group wars, conflicts among the drug structures, wars against the police and the police against drug pushers. All are within drug wars throughout the world. The truth is the whole world is in a mess with wars.

However, Jesus reminds believers that they should not become alarmed because such prophecies must come to fulfillment. When these things come to pass, remember that he said that the end is still to come, meaning many more events will take place before the end of time. Jesus Christ calls these wars, rumors of wars, earthquakes, famines, and pestilences, *the beginning of birth pains* and are signs of the "End Times." The many stories of the news media in the previous chapter, show evidence that these events that Jesus Christ predicted are taking place.

As readers move into the predictions of the apostles in the following chapter, the apostle Paul and others predict similar events to take place. Jesus gave the apostle John the blueprint of the prophetic eventful coming-signs of the *End Time.* The *End Time* events will be *Signs* occurring in the sky, on the land, in the waters, and with the weather. People will see these as strange and unusual, but they will only be the beginning of sorrow and pain that will come upon the world, Jesus predicted.

Unusual Signs with Disease and Pain in the World

Jesus Christ's predictions are *'Signs'* that much pain would consume the world before he comes back. For example, wars bring pain. Wars kill, wound, and destroy families. There is also the pain of leaving one's home or paying one's life-savings to escape a war-ravaged land or losing loved ones. Worst of all, after reaching another nation, it is painful when the nation rejects the refugees and refuses to give help. Connected with the pain of war and nations revolting against each other, Jesus Christ predicted there will be famines and earthquakes.

The pain of famine is happening today in some countries. Famine brings pain and more pain, especially in seeing children dying because of the lack of food. There is also the pain of earthquakes. Earthquakes will take place in various places around the world, resulting in much pain for communities and people. Again, the pain of death is clear where earthquakes are taking place. Earthquakes destroy the economics of many countries, and various kinds of diseases and pestilence follow.

Today, science is discovering many new diseases, but there are no cures for them. Viruses attack people around the world. Diseases will become a pest to the people of the world. For example, the news media and science are proclaiming that the manifestation of pestilence is already among us. Note, as of January 30, 2020:

> "If things go on like this, we will all be doomed," said a woman identifying herself as Xiaoxi, in an interview over the weekend with the *South China Morning Post*.
> "People just keep dying; no one is taking care of their bodies."
> "Xiaoxi's husband is among the 7,600 people across China who are confirmed to be infected by a new pneumonia-like illness called the 2019-nCoV, or Wuhan Coronavirus."
> "some 5 million people are believed to have left Wuhan for other parts of China and the world.
> "The true cause of the virus may still be unknown, that the strain that infects humans emerged earlier than official reports say, and that it then remained undetected for an unknown duration. (The Trumpet, Jeremiah Jacques, (2020).
> https://www.thetrumpet.com/21859-the-wuhan-coronavirus-and-the-bibles-prophesied-disease-pandemics

Since these announcements, as of February 19, 2020, 14,000 people have been infected and 2000 have died from the virus. Each day the pestilence of the Coronavirus is making new headlines. It now appears to be spreading from China to other areas of the world. So far, the photo below shows the spread of the virus. Note the stats and places identified.

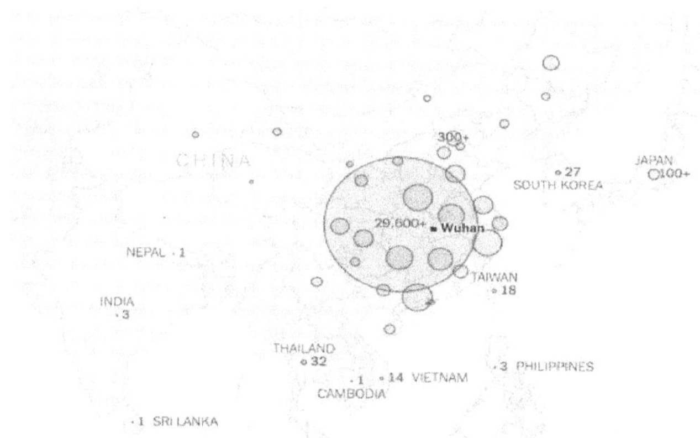

<u>Map above showing the spread of the Coronavirus as of 2/10/2020</u>

Not only in Jesus Christ's predictions about "pestilences' will come upon the people of the world before His return to the earth, but also in Revelation 6:8, deaths will destroy many people by various pestilences. Death will kill one-fourth of the world's population with weapons of war, hunger, and other means such as pestilences, of humans and animals. Also, in many chapters of Revelation, the apostle John predicts that creatures will come upon the earth and will sting people. The pain will last for several months. Also, in Revelation 16:10-11, the pestilence of sores will come upon the people of the earth. John predicts many other forms of pestilence to come upon people in the End Time. The Bible Epistles chapter presents and explains many of the predictions about End Time pestilences that will come upon the earth.

*Unusual Signs with the Great Tribulation Will Come upon the **Earth***

As Jesus continued sharing these prophecies with his disciples that day on the Mount of Olives, he also predicted even greater distress. He said, **"For then shall be great tribulation, such as was not since the beginning of the world to this time, no, nor ever shall be"**

(Matthew 24:21). Although there will be the pain of wars, famines, earthquakes, and diseases, these horrible events are prior to a greater event that will engulf the world. This event is "The Great Tribulation." The Bible predicts horrifying eventful signs will occur which people have never witnessed before.

It is important to note that Jesus Christ's coming back to the earth will be after the seven-year great tribulation that will come upon the world. The great tribulation has not started yet. Jesus Christ has not yet taken His Believers from the earth. Also, He has not yet taken the scroll from the Father's hand around the throne, as stated in Revelation 4. Jesus Christ has not yet pronounced the beginning of the great tribulation on the earth. The world does not yet have an Antichrist ruler. The world has not yet had a religious leader called the false prophet. The world has not yet had a political leader known as the beast. The world does not yet have a one-world currency. It does not yet operate the financial systems by a mark on the forehead and on the hand. But we must admit, the signs that are taking place today, the previous events discussed, shows that the Great Tribulation is soon to take place.

Unusual Signs with the Plight of the Israeli People

Jesus Christ referred to a former prophecy from Daniel the prophet and stated:

> When you, therefore, shall see the abomination of desolation, spoken of by Daniel the prophet, stand in the holy place, (whoso reads, let him understand:) Then let them who are in Judaea flee into the mountains: Let him who is on the housetop not come down to take anything out of his house: Neither let him who is in the field return back to take his clothes. And woe unto them that are with child, and to them that give suck in those days! But pray ye that your flight is not in the winter, nor on the sabbath day: For then shall be great tribulation, such as was not since the beginning of the world to this time, no, nor ever shall be.
>
> And except those days should be shortened, there should no flesh be saved: but for the elect's sake, those days shall be shortened. Then if any man shall say unto you, Lo here is Christ, or there; believe it not. For there shall arise false Christs, and false prophets, and shall shew great signs and wonders; insomuch that,

> if it were possible, they shall deceive the very elect. Behold, I have told you before.
> Wherefore if they shall say unto you, Behold, he is in the desert; go not forth: behold, he is in the secret chambers; believe it not. For as the lightning cometh out of the east, and shineth even unto the west; so, shall also the coming of the Son of man be. For wherever the carcass is, there will be the eagles gathered together. (Matthew 24:15–28)

The *primary interpretation* of the entire book of Matthew is about the nation of Israel. So, in this passage, Jesus gave the disciples the answer to the first question they asked—when will these things happen? Here, he gives them a two-fold answer.

First, He predicted the destruction of Jerusalem, as Daniel prophesied.

Second, He describes the havoc and distress and relates some of it to events of the Great Tribulation that will come upon the earth and how the Israeli people will suffer during that period.

Third, He also addressed the fact that the Israeli people will have problems with a coming Messiah. They rejected Jesus and are still looking for a coming Messiah. Many false prophets will come and will perform miracles like those that would be present in the coming great tribulation. However, he mentions his literal Second Coming as the real Messiah, and he will set up his kingdom. This is basically a second chance for the Israeli people.

Fourth, prior to Jesus' start of His ministry, John the Baptist preached to the Israeli people about the gospel of the kingdom. However, they rejected Jesus Christ as the Messiah and therefore, rejected the promised- kingdom. The rejection caused God to lay them aside. So, God started the "calling out" of any person who would accept Jesus Christ as Savior, thus, forming His Church.

About this shift God made, the apostle John gives the clarificationin the Gospel of John 1:6-18). John explains how (1) John the Baptist came from God and was a witness of the Light, Jesus Christ. (2) Jesus Christ was to lighten all people who will believe in Him. (3) He is the One who made the world, yet the world was and still is ignorant of who He is. (4) He came to the Israeli people, but they rejected Him as the promised Messiah. (5) Therefore, God turned to anybody who will believe in His Son Jesus Christ, God gives to that person everlasting life. (6) To get this kind of life, a person must be born of God. (7) The Person Jesus Christ is the Only One who can give this new birth, for He is the "Only Begotten of the Father.

Fifth, we must note here that the preaching of the kingdom by John the Baptist differed from the preaching of the gospel of Jesus Christ. The message of the kingdom is, "Repent for the kingdom of God is at hand." The message of the gospel of Jesus Christ is "Trust in the Lord Jesus Christ as your Savior and you will have eternal life." We can also sum this up as "Anyone who Believes in the Lord Jesus Christ, He will save." The Jewish people did not accept their Messiah, and they are still waiting for his coming. They will continue to have tribulations until he returns the second time.

Sixth, before Jesus Christ returns to earth with his bride, 144,000 witnesses, all Israeli men, all virgins, He will seal. They will then preach the kingdom gospel again in the Great Tribulation. Revelation 7:9 gives the results of their preaching. A great multitude that no one could count, from every nation, tribe, people, and language, will stand before the throne and before the Lamb. They will wear white robes and will hold palm branches in their hands. God will again turn to the Israeli people during the Beast/Antichrist dictatorship in the Great Tribulation. They will again hear the preaching of the Kingdom Gospel by the 144,000 young Israeli men.

Those who have turned to Jesus Christ because of the preaching of the kingdom gospel will go into the millennial kingdom that Jesus Christ will set up on his return to the earth. (see Revelation 7:1–10)

An Unusual Sign of Warning—Learn the Lesson of the Fig Tree about Jesus Coming Back to the ***Earth***

Jesus gave a warning to his disciples about the times related to his coming back to the earth. He told them that they needed to learn a lesson from the fig tree. This interpretation of his predictions is in Matthew 24:32–36. According to this scripture,

First, the fig tree gets twigs that are tender, followed by leaves. This is a sign that the "summer" of Jesus's coming. "

Second, as the tender twigs emerge from the fig tree, and the leaves follow, everybody knows that this sign means summer is near.

Third, therefore, the lesson is, when people see (1) the signs of *famine, wars, and rumors of wars*, (2) *deceivers of God's truth,* (3) *believers turning away from God*, and (4) *persecution of believers*, not only by the world but by believers themselves, people need to understand that the *summer of Christ's return is near*.

Fourth, in addition, when people are living like those in Noah's day – parties and not giving any attention to God, is also another sign that the '*summer'* of his return is near.

Fifth, the reason he gave this lesson, His love is so great for humankind that he does not want to see them in a lost condition when he comes. He knew that some people would not listen to the gospel and be ready to go with him. This is not because of him but because of their own choice. He made this clear in his prophecies when he said that two men will be in the field; a power will take one and the other left. Two women using the hand mill; the power will take one and the other left (Matthew 24:36–41). Because Jesus does not want to leave anyone behind, he added in his prophecy the following warning as he spoke with his disciples:

> Watch therefore: for ye know not what hour your Lord doth come. But know this, that if the goodman of the house had known in what watch the thief would come, he would have watched, and would not have suffered his house to be broken up. Therefore, be ye also ready: for in such an hour as ye think not the Son of man cometh. (Matthew 24:42–44)

Jesus stated that no person knows when He will return to the earth. Only the Father in heaven knows the time. Many people have tried predicting it but only make themselves liars. This coming of Jesus Christ back to the earth is so secretive that he described it as being like the strategy and plans of a thief going to a person's house to break in and steal. However, he gave signs and the parable of the fig tree.

Sixth, note the concluding lesson he taught his disciples. They were to watch because they did not know when the Lord would come. Jesus Christ had come to the Israeli people, but probably because they were not watching, they failed to see who He was because He came unexpectedly.

This second story speaks of a caretaker who his employer left him to take care of the servants. His employer left instructions that if he does a good job he will give him a reward. However, if he does not take care of the servants, he will condemn him to a place where the hypocrites weep and gnash their teeth.

The lesson for the believers of today is that Jesus Christ is the Savior of the believers. However, he has gone away, as he said, to prepare a place and will come back for the believers so they may be with him forever. However, in the meantime, the believers of Jesus Christ are not to just wait for him; they are to engage in the works that he has given them to do. Believers are the light and salt of the earth, and they must share their presence and the message of Jesus Christ with the world. In fact, in one of Jesus Christ's teachings to the multitudes, he stated:

> Neither do men light a candle, and put it under a bushel, but on a candlestick; and it giveth light unto all that are in the house. Let your light so shine before men, that they may see your good works, and glorify your Father which is in heaven. (Matthew 5:15–16)

Also, Matthew 28:19–20 is the last promise Jesus Christ gave the disciples before He ascended into the clouds. Note what he told them, "Go ye therefore, and teach all nations, baptizing them in the name of the Father, and of the Son, and of the Holy Ghost: Teaching them to observe all things I have commanded you: and, lo, I am with you always, even unto the end of the world. Amen."

Jesus Christ Predicts Unusual Signs in the Waters and with the Weather

Within Jesus Christ's predictions on the weather, he gave a similarity between the days of Noah and the time of his coming back to the earth. For example, he stated**, "And knew not until the flood came, and took them all away; so, shall it be when the Son of Man comes"** (Matthew 24:39). The people of Noah's day were aware of the flood coming because they heard the preaching and message of God's Messenger for 120 years. They heard Noah, but they perceived Noah's message to be *"Fake News."* They rejected the message and therefore, perished. See, to those people, a flood would be strange and unusual. They had never seen rain, so they did not accept Noah's message about God's judgment with a flood. To many people today, the strange and unusual signs Jesus Christ predicted that will occur before He returns to the earth, many people do not take seriously.

Jesus Christ predicted that just as the people in Noah's day did not believe in the *'coming rain'* people in the *End Times* will not believe that the *weather-events* such as *earthquakes, lightning and thunderstorms,* and *famine, are signs* of His coming. Yet, He wants people to understand that *weather events* throughout the Bible have a connection to the Judgment of God. And, in this case, He likened the events to the *End Times.* Note the *End Times'* events Jesus Christ predicts to come upon the sky and land because of the weather. In Matthew 24:7b He states, **"there shall be famines, and pestilences, and earthquakes, in diverse places."**

Today, global climate change in the weather is causing the soils to become unproductive, there are shortages of food, water consumption is dwindling, pests are increasing, thunderstorms, earthquakes, and such kinds are occurring in the world. Many people are looking for answers. Let us investigate famines and earthquakes.

The Sign of Unusual ***Famines*** *in Various Parts of the Earth Before He Returns*

Jesus Christ's prediction about famines in various parts of the earth, for years after He left the earth is taking place. However, today they are occurring in increasing numbers. According to science, famine is a product of Global Climate Change. The heating of the earth destroys the soil through hardening. Therefore, when it rains, the water runs off, causes flooding, and never soaks into the ground. Therefore, the hardened soils make it impossible for the crops to grow. If there are no crops it means that there will be short supplies of food. Also, if the process of global warming continues, the changes in the climate will make famines worse.

We can also add that the wars and rumors of wars that Jesus predicted to come in the *End Times*, also influence famine. As wars are going on, *famine* follows. There will be shortages of food, and when there are wars and no food, people flee, looking for refuge. They become refugees in other lands and sometimes the authorities refuse to accept them. Today because of wars and uprisings in the world, Jesus Christ's predictions are manifesting already. Could it be that the Great Tribulation is near? If so, it means that the Rapture of the Believers is even closer. The Great Tribulation will not begin until Jesus Christ first takes all the Believers from the earth.

The Sign of Unusual ***Earthquakes*** *in Various Parts of the Earth Before He Returns*

The prophecy of *earthquakes* has a connection with *famines*. Since Jesus Christ's predictions to his disciples on the Mount of Olives, many earthquakes have taken place around the world. Again, as seen in previous chapters of this book, earthquakes are occurring in greater frequency. The findings stated that earthquakes and their devastation rob countries of food and water. They destroy buildings, industries, and farming. The destruction then leads to famine.

Remember—Jesus Christ has not returned to the earth yet. Also, His predictions are about the *End Times.*

Concluding Statement

First, the following events which Jesus Christ predicted will continue to occur until He comes back to the Mount of Olives, outside the city of Jerusalem, on His return to the earth through the *sky* (Zechariah 14:4 "And his feet shall stand in that day upon the Mount of Olives, which is before Jerusalem on the east…").

Second, on the *lands of the earth,* (1) the *deception* of leaders, politicians, religious leaders will continue. (2) Many *'false Christs* and *false prophets'* (Matthew 24:11, 24; Mark 13:22; Luke 6:26; 2 Corinthians 11:13;

1 Peter 2:1; 1 John 2:18; 4:1; 2 John 1:7; Revelation 16:13; 19:20; 20:10;) will continue to show up in the *End Times*. (3) Many believers of Jesus Christ will turn away and will even deny each other. (4) On earth, there will continue to be wars and rumors of wars, famines, and pestilences. (5) Diseases without cures will continue to plague

humankind.

Third, the ***weather*** will continue to produce strange and unusual events such as (1) *bolts of lightning, thunderings, earthquakes/tsunamis, and famines.* (2) The Nation of Israel will continue to be under the attack of the Gentile nations. (3) The greatest attack will take place during the Great Tribulation when their greatest enemy, the Beast/Antichrist will attack them and desecrate their place of worship. (4) The *Great Tribulation* must come upon the earth someday soon. Do you know why? Jesus Christ predicts that these strange and unusual signs will take place in the sky, on the land, in the waters, and with the weather-Remember, he is God in the flesh. He knows everything.

Fourth, with all the strange and unusual signs taking place on earth, (1) science will blame Global Climate Change. (2) Some people will say it is mother nature while others say God is speaking. (3) Then, there will be those who will believe nothing. (4) Yet, the Bible is clear about the predictions of Jesus Christ and the *coming 'End Times'.* However, Jesus Christ predicted that many will not believe. They will be like those of Noah's time regarding the flood.

Fifth, Jesus Christ linked the *deception* in the *'End Times'* to the flood on the earth in Noah's day. So, the signs He predicted would come in the sky, on the land, in the waters, and with the weather, people will not believe. Many believers will become weak and will turn from the faith. Some will even turn against each other and deliver others up for destruction.

Sixth, talking about violence, nations will rise against nations, and there will be wars all over the world. This violence of people and the wars will cause much pain among the people. There will be an increase of diseases and pestilence. The people of Israel will undergo much torture, harassment, and hate from the Gentile nations right up to the Great Tribulation. However, as the Great Tribulation ends Jesus will return to the Mount of Olives from the sky.

Seventh, Jesus warned people to watch because His coming will be a surprise, just as Noah's flood was. He established that his coming back to earth will be an unusual one. He did not tell the timing of his return to the earth, because only the Father knows such information. He said he will enter through the sky, and he predicted that there will be signs in the heavens that will influence the earth. The sun, moon, and stars will not give light to the earth. He likened these signs to the fig tree, and it signals to people that the summer is coming. Therefore, we can conclude that the signs that will take place with the weather, and their effects are a clear sign that He will return very soon.

Eighth, the prophecies of Jesus Christ are clear indications that He is coming soon. Many of the Old Testament prophets and the New Testament apostles predicted similar 'End Time' signs concerning the return of the King of kings and Lord of lords to the earth. The Old Testament prophets did not know this 'Person' who is coming as Jesus Christ. They only knew Him as the Messiah. However, the New Testament reveals Jesus Christ as the Messiah. Jesus Christ warned the Israeli people and all other people to watch and be ready for His return. Also, the timeline of His return, He gave it in the Fig Tree Parable. The message Jesus is giving informs us that the tender twigs and the leaves coming from the fig tree show that **"summer is near."** Also, when people see all the signs Jesus Christ predicts for the "End Time" – wars, famines, earthquakes, pestilences, deception of religious and political leaders, lack of love, a special hate for the Israeli people, they must know that the **"End Time" is near, right at the door"** (see Matthew 24"32-33).

The signs that Jesus Christ predicted seem to already start on the earth. Wars, news about wars, uprisings, famines, pestilences, earthquakes, and other storms, are all occurring today. Jesus Christ's predictions about "End Time" signs are like the events scientists are calling events associated with Global Warming and Climate Change. Does this mean that global warming/climate change and the prophecies of the Bible have common ground? If so, it would mean that "summer" or the "end" is near. And Jesus Christ is coming back to the earth very soon. Therefore, people should not be like the people of Noah's day. We should all use wisdom, listen to the messages of the prophecies of the Holy Bible, and accept the preparation which is through the blood of Jesus Christ. There are soon coming *imminent events. Signs or events* such as (1) the "snatching away" of the believers to the skies, (2) the escape from the seven years of Great Tribulation on the earth, and (3), the actual return of the Messiah, Jesus Christ to the earth by Mt. Olives.

Chapter 16

MARK'S GOSPEL PRESENTS JESUS CHRIST'S PREDICTIONS OF UNUSUAL 'END TIME' SIGNS

The Apostle Mark

The apostle Mark also shared the ministry of Jesus Christ on earth. He was a witness to Jesus's teachings, miracles, death, and resurrection from the dead. Mark was one of Christ's twelve disciples who was constantly with him. He was one of the selected twelve apostles. History shows that he was born in Cyrene, Libya. The Holy Bible records his missionary work with the apostle Barnabas, who was his uncle, and the apostle Paul. Although his youth hindered him from the hard travels of the missionary journey with Barnabas and Paul, the apostle Paul did not want to take him on another trip. Later he grew into a giant of an apostle.

The apostle Paul realized that this brother had grown in the Lord. In fact, when Paul was old and felt that he had fought a good fight, had kept the faith, and had finished his course, he asked that the apostle Mark come to him to assist with his ministry. "Only Luke is with me. Take Mark and bring him with thee: for he is profitable to me for the ministry" (2 Timothy 4:11). Mark was one of the chosen apostles whom the Holy Spirit used to write the Gospel of Saint Mark in the Holy Bible. The apostle Mark saw and referred to Jesus Christ as the Son of Man. Mark saw Jesus Christ as a Savior who dedicated his life to the service and salvation of humankind. In the Gospel of Saint Mark, the apostle referred to strange and unusual signs that will take place in the sky, on land, in the waters, and with the weather.

Jesus Christ Predictions of Unusual Signs/Events in the Sky and

on Land

Jesus Christ's prophecies of unusual signs/events in the sky, on land, in the waters, and with the weather, the apostle Mark recorded in the Gospel of Mark. Mark describes the signs Jesus Christ predicts this way:

> But take ye heed: behold, I have foretold you all things. But in those days, after that tribulation, the sun shall be darkened, and the moon shall not give her light, And the stars of heaven shall fall, and the powers that are in heaven shall be shaken. And then shall they see the Son of man coming in the clouds with great power and glory. And then shall he send his angels and shall gather together his elect from the four winds, from the uttermost parts of the earth to the uttermost part of heaven.
>
> Now learn a parable of the fig tree; When her branch is yet tender, and putteth forth leaves, ye know that summer is near: So ye in like manner when ye shall see these things come to pass, know that it is near, even at the doors. Verily I say unto you, that this generation shall not pass, till all these things are done. Heaven and earth shall pass away: but my words shall not pass away. But of that day and that hour knowest no man, no, not the angels in heaven, nor the Son, but the Father.
>
> Take ye heed, watch and pray: for ye know not when the time is. For the Son of man is as a man taking a far journey, who left his house, and gave authority to his servants, and to every man his work, and commanded the porter to watch. Watch ye therefore: for ye know not when the master of the house cometh, at even, or at midnight, or at the cockcrowing, or in the morning: Lest coming suddenly he finds you sleeping. And what I say unto you I say unto all, Watch. (Mark 13:23–37)

The Predictions of Unusual Signs/Events in the ***Sky***

The apostle Mark's description of this prophecy he directs solely to people. He describes the signs where the "sun and the moon will not give any light to the earth. Also, another sign will take place with the stars. They will fall from the sky and the powers of the heavens will shake." Whatever is holding the heavens together will shake when Jesus Christ is on His way to the earth. No doubt this shaking in the sky also will have to shake the land. Activities such as these will also affect the people on earth. Can you imagine the amount of fear and confusion on the earth when Christ returns?

The Predictions of Unusual Parables about the ***Earth***

The apostle Mark records the two examples Jesus used to illustrate the importance of people being ready for His coming. The first was the case of the fig tree. The second was the case of the man who had gone on a long journey but will return to his house one day. These two cases show that Jesus Christ is that man who has gone on a long journey and will come back. The fact is his servants need to take heed and watch because no one knows the hour or the day when this man will return. However, he shows that as with the fig tree, every person understands when this tree buds, summer is near. Therefore, the signs that people see today are indicators of the "End Time." The apostle wanted humankind to be ready for the return of Jesus Christ.

Chapter 17

Luke's Gospel Presents Jesus Christ's Predictions of Unusual 'End Time' Signs

The Apostle Luke

Luke was also one of the twelve apostles close to the ministry of Jesus Christ. People believe that the apostle was a medical doctor. In fact, Paul mentioned in 2 Timothy that Luke was with him. The apostle Luke was present with the Lord Jesus Christ on the Mount of Olives when Jesus explained the coming signs people will see in the skies, on land, in the waters, and with the weather, before he comes back to the earth. Whatever the rest of his life was, it is evident that he was an apostle of Jesus Christ and the one who received from God, through the Holy Spirit, the prophecies of Jesus Christ that will take place in the End Time.

Jesus Christ Predicts Unusual Signs/Events in the Sky, on Land in the Waters, and with the Weather

The Gospel of Luke has similar recordings of the predictions of Jesus Christ like those of the apostle Matthew. Most of his Gospel 'End Time' prophecies are in Luke 21:7–36. The prophetic predictions were presented when Jesus' disciples asked him this question, **"Master, but when shall these things be? and what sign will there be when these things shall come to pass?"** His answer to them, He gave several prophetic signs/events that will take place before His return to the earth. Note the outstanding points in Luke 21:

1. Deceivers are coming and will confess to being Jesus Christ.
2. Wars among nations and kingdoms are coming.
3. Earthquakes, famines, and pestilences are coming to the

world. They will produce fearful sights and great signs coming from heaven.

4. The enemies of Jesus Christ will capture, persecute, deliver, put into prisons, His followers because of their testimony and their belief in His Name.
5. Parents, brothers, other family members, and friends will betray the believers of Jesus Christ and will cause some to lose their lives.
6. All people will hate the followers of Jesus Christ.
7. The Gentile armies will destroy the City of Jerusalem and will kill the people. The hate which the gentiles have for the Israeli people will continue until the end of the "Times of the Gentiles."
8. People will see the "End of Time Signs" in activities of the sun, moon, and the stars. The signs will cause distress to nations and perplexed people.
9. People will also witness "End Time Signs" in the sea. The roaring of the waves will add fear in them, and many will have heart failures.
10. The power of God will shake when the people of the earth witness the Son of Man coming in a cloud with power and great glory.
11. When the above signs begin, people must look up, lift up their heads because their redemption is very close.
12. The sign for these signs/events coming to an end is in Jesus' Fig Tree Parable. For example, the fig tree coming with new leaves shows that summer is near. So, when people see deception, wars, killing, hate, signs in the sky, on the land, in the waters, and with the weather, it means that "Jesus Christ's *Summer of His Coming* is near."
13. So, rather than carousing, drunkenness, and anxiety of life, watch, pray, and be ready to escape the coming eventful signs. (teachings from Luke 21:28–36).

From the above prophetic predictions of Jesus Christ, there will be strange and unusual signs taking place in the sky, on the land, in the waters, and with the weather.

The Predictions of Unusual Signs/Events in the ***Sky***

First, in the sky, there will be signs in the sun, moon, and stars. In the above prophecy, the sun, moon, and stars will have signs.

Second, Jesus did not say what kind of signs, but He predicted what the *big coming sign* will be. All the people of the earth will see Him, the Son of Man coming on a cloud with power and great glory. According to the prophet Zechariah of the Old Testament, he predicts that the Messiah will return from the sky to the Mount of Olives. Well, it was from the Mount of Olives that He predicted all the 'End Time' signs. It is on the Mount of Olives where we find the location of Golgotha or Calvary, and also where the Roman soldiers crucified Him. It was at the Mount of Olives where He ascended to heaven and is at the right hand of the Father. It will be from the Mount of Olives where He will return with the Saints and the armies of heaven at His Second Coming to the earth. See Acts 1:4–11, where the angels told his disciples that**, "This same Jesus, who has been taken from you into heaven, will come back in the same way you have seen Him go into heaven."**

Third, on that day when the Son of Man is coming back to the Mount of Olives, there will be flashes and lights in the sky. These lights will be all over the sky, from one end of the heavens to the other.

Please remember, (1) this prophecy about the coming of Jesus Christ to the earth is NOT THE RAPTURE OF THE BELIEVERS OF JESUS

CHRIST. (2) It refers to the SECOND COMING OF JESUS CHRIST to the Mount of Olives. (3), Before this event, Luke records Jesus' coming back, and other signs Jesus told them that occur.

Fourth, the apostle Luke records another prophecy of Jesus Christ which clarifies that before He comes back to the earth, He will first suffer. Note, **"But first must he suffer many things, and be rejected by this generation"** (Luke 17:25). This prediction associates the Jewish nation's rejection of Jesus Christ as Messiah and the building of His church, the called- out assembly. Such work of establishing this church of Jesus Christ is by Jesus Christ alone (see Matthew 16:13–20).

Predictions of Unusual Signs/Events on the ***Land/Earth***

Luke's account sums up the predictions of Jesus Christ about the land with (1) Signs of Deception among the people of the world, (2) Signs of Wars among the people of the earth, (3) Signs of Earthquakes, Famine, and Pestilences, Fearful Events, and Great Signs from Heaven, (4) The Sign of People Living Lifestyles Like Those of Lot's Days in Sodom, will give Evidence to the 'End Time'.

The Sign of Deception Among the People of the World

The apostle Luke records Jesus's prediction about the sign of deception that will be clear in the "End Time." In the End Time, false messiahs and leaders will claim to be Jesus Christ. They will also tell people that the end is near. Jesus Christ warns His Believers not to follow such people. Luke encourages the believers with the words of Jesus Christ which He shared with the disciples on the Mount of Olives. Note, "And he said, Take heed that such people will not deceive you: for many shall come in my name, saying, I am Christ; and the time draws near: go ye not therefore after them" (Luke 21:8).

The Sign of Wars among the Nations on ***Earth***

Luke also records prophecies of other eventful signs that will take place on the earth before Jesus Christ returns to the earth.

First, Luke predicts Jesus' prophecy about uprisings and wars. In Luke 21:9-11, the apostle Luke records the above prophecy about coming uprisings and wars. Since Jesus Christ gave this prophecy and the recording of the apostle Luke, there have been many wars and conflicts.

Second, of the conflicts that Jesus Christ predicted, Luke's recordings of Him also predict that people who follow Jesus Christ will have many problems on earth. Jesus' followers will become targets of hate, betrayals, persecutions, and even deaths. Such conditions of betrayal will be by people who are very close to the followers of Jesus. For example, they will be parents, brethren, and friends (Luke 21:16-19)

Third, the apostle Luke recorded the prophecy of Jesus Christ about the coming *Great Tribulation* on the earth, and how it will affect the Israeli people (see Luke 21:20–24). The apostle brings out Jesus' predictions about the nation of Israel.

A. The Gentiles will besiege the city of Jerusalem. The people will run to the mountains for refuge. This apparent judgment is because of the rejection of Jehovah's word and standards.

B. The Jewish people will be prisoners to all the nations. The nation is not a prisoner in all countries today, but many nations still have a hate for the Israeli people. This hatred appears to be over *land ownership*. Remember they do not have all the land that God promised to them in the *land covenant*, sometimes referred to as the *Palestinian covenant.* The covenant states:

> In the same day, the LORD made a covenant with Abram, saying, Unto thy seed have I given this land, from the river of Egypt unto the great river, the river Euphrates: The Kenites, and the Kenizzites, and the Kadmonites, And the Hittites, and the Perizzites, and the Rephaims, And the Amorites, and the Canaanites, and the Girgashites, and the Jebusites. (Genesis 15:18–21)

C. The land where many tribal groups of people lived, God decided it was His land, so he gave the land to the nation of Israel. Note what God said:

> And I will establish my covenant between me and thee and thy seed after thee in their generations for an everlasting covenant, to be a God unto thee, and thy seed after thee. And I will give unto thee, and thy seed after thee, the land wherein thou art a stranger, all the land of Canaan, for an everlasting possession; and I will be their God. (Genesis 17:7–8)

D. God established the land covenant between the nation of Israel and Himself before the witness, Abraham. The covenant is generational, with an everlasting possession of the land. The measurement of the land in the covenant is from the river of Egypt to the great river, the river Euphrates. Such an area covers where other nations live today. But remember, the land is an agreement between God and the Israeli people. Therefore, they must one day have it (see Genesis 26:2–4). Also, God appeared to Jacob and reminded him of the land covenant he'd formerly made with Abraham (see Genesis 28:10– 17).

When Jesus Christ returns and touches down on the Mount of Olives, the Israeli people will accept Jesus Christ as their Messiah. At such a time, they will receive all the promised blessings of God.

The Signs of Earthquakes, Famine, and Pestilences, Fearful Events, and Great Signs from Heaven

First, the apostle Luke continued to present Jesus' prophecy about coming signs/events with the weather. For example, he records in Luke 21:11, **"And great earthquakes shall be in diverse places, and famines, and pestilences; and fearful sights and great signs shall there be from heaven."** Today, earthquakes are increasing in frequency and strength all over the world. These storms not only destroy the land and kill people but sometimes, they create powerful tsunamis, which destroy everything in their paths. Luke's Gospel predicts these types of storms to take place in the *'End Times'* before Jesus Comes back.

Second, besides earthquakes and tsunamis, Luke's account states where Jesus predicts *famines* and *pestilence* to overtake the world. There are people in this world who do not have food to eat. Wars and uprisings have increased the potential of famine. Many countries are not producing food as they once did. Scientists suggest that global warming/climate change is responsible for much of the decrease in the production of food. The changes in the climate have caused this problem, but the prophecy has predicted that it will come to pass.

In addition to famines, it is predicted that there will be pestilences. The word "pestilence" in the Greek Language is *Loimoi* written in English. This word means pestilence or plague as stated in Matthew 14:7 and Luke 21:11. However, *Loimoi* also means that people may see a person as a pestilence. Note, **"For we found this man, a pestilent fellow, and a mover of sedition among all the Jews throughout the world, and a ringleader of the sect of the Nazarenes."** (Acts 24:5) A man was here preaching, and others labeled him as 'a pestilence,' meaning, he is destructive or pernicious. The teaching shows that in the End Time, not only mosquitoes and other pests people will classify as plagues or pestilences, but even other people.

The Sign of People Living Lifestyles Like Those of Lot's Days in Sodom will give Evidence to the 'End Time'

With all the strange and unusual activities, people will still ignore God's Word. They will be like the people in Noah's and Lot's days. The apostle Luke records Jesus' prediction in Luke 17:26–35

First, in these prophetic scriptures, the apostle Luke described how Jesus Christ predicts the condition of people's living before He comes back to the earth. The example of Lot and Noah from the Old Testament illustrates the type of living conditions. He mentioned that the people in Lot's day did not listen to the warnings from the strange men who visited and warned them of coming judgment on Sodom and Gomorrah. However, the day Lot left Sodom, God rained fire and brimstone from heaven and destroyed them all.

Second, Luke's account used this case study to show that one *sign* of the coming of the Lord Jesus Christ will be some people living like those of Sodom and Gomorrah. Luke saw this choice of life as one destined for destruction. The people of Sodom and Gomorrah were more concerned about their ways of living. In fact, the men of those cities refused to take Lot's virgin daughters to themselves. They wanted to have intimate relations with the visiting men. Their sexual orientations appeared to be for the same sex. The prophecy shows that before Jesus Christ comes back to earth, people will live in ways that are like the people of Sodom and Gomorrah lived.

Third, comparing the prophecies of Jesus Christ as presented by the apostles Matthew, Jude, and Luke, all three speak to signs of the coming of Jesus Christ to this earth. They all predict that peoples' lifestyle during the *'End Times'* will be like those of Sodom and Gomorrah and like those of Lot's and Noah's days. Meaning, people will engage in violence and sexual immorality leading up to the return of Jesus Christ's return to the earth.

The Sign of Unusual Signs/Events in the Waters and with the Weather

First, the apostle records that there also will be unique signs or events happening in the waters. For example, in Luke 21:25 the prophecy states, **"And there shall be signs in the sun, and in the moon, and in the stars; and upon the earth distress of nations, with perplexity; the sea and the waves roaring;"** The roaring of the sea and the surging of the waves will have a great effect on people throughout the world. The description here, regarding the activity of the seas, tells us the people will become bewildered, perplexed, and troubled; it will become a tormenting force for people. For example, the roaring of the sea will cause people to experience anguish and perplexity. There will be constant inclement weather; storms also will occur. The problem will be a global one.

Second, problems will compile, and the eventful signs that take place will bring about dismay among all nations. Because of this torment from the activities in the sky and the noise of the storms on the earth, people will faint from fear and anxiety. After experiencing such tormented storms, the sky will begin shaking, and **"Men's hearts failing them for fear, and for looking after those things which are coming on the earth: for the powers of heaven shall be shaken"** (Luke 21:26). It is apparent that the world will experience a super-global climate change effect.

Third, the most unusual event will be when people see the Son of Man coming on a cloud with power and great glory. There already are signs of unusual activity in the sky. Meteorites are falling to the ground; unusual snowstorms are occurring. There have been temperature changes, and the earth and oceans are becoming warmer. Scientists say the sun is closer to the earth's surface. Global warming/climate change is clear in our day, but there is more. Who is conducting this phenomenon? Is global warming/climate change? According to the prophecies, God is still in control, and these predictions must come to pass.

The Apostle Luke's Gospel Assures the fulfillment of Jesus Christ's Predictions

First, Luke believed that Jesus' prophecies will definitely, without a doubt, come to pass. In his writings, he stated what Jesus Christ had predicted on the Mount of Olives. For example, in Luke 21:28–36, the prophecy predicts that when earthquakes, pestilences, famines, wars, conflicts, and hate are manifesting on the earth, these will be signs of the

End Time.

Second, when people see the *'End Times'* signs beginning to take place, Luke warns:

1. People must look up and lift up their heads, because "redemption is close at hand." (see Luke 21:29–31)
2. People must pay attention to these signs because Christ is about to return to the earth. (see Luke 21:32–33)
3. People must take heed. Rather than letting carousing, drunkenness, and the anxieties of life weigh down their lives, they should look up and never allow these fleshly events to entrap them like a mouse in a trap. (see Luke 21:34–35)

Third, the apostle Luke records that the strange and unusual activities in *the sky,* on *land,* in *the waters*, and with *the weather* will come to pass.

Section Four Concluding Remarks

First, all three gospels reiterated the prophecies of Jesus Christ in a harmony that shows the connectivity and agreement of God's Word. For example, they all (1) identified Jesus' prophecies of *End Time Signs* of wars, rumors of wars, earthquakes around the world, famine, and pestilences. (2) All Gospels stated Jesus' predictions of coming deception that will overtake the world. (3) All Gospels share Jesus' warnings of preparedness for the coming of the Messiah. (4) They all stated that Jesus' predicted *End Time Signs* to occur in the sky, on land, in the waters and with the weather.

Second, the Gospels presented many of the *End Time Signs* that Jesus Christ predicted will take place before He returns to the earth. In addition to these *End Time* prophecies from the Gospels, the Epistles of the New Testament also have many predictions of the End Time Signs to share. Let us now read what the apostles Peter, Paul, and John predicted in Section Five.

SECTION FIVE

The New Testament Epistles Predict Unusual 'End Time' SIGNS

The 'Epistles' of Peter, Paul and John Predictions about Coming 'End Time' Unusual Signs/Events. These letters were written to the Church of Jesus Christ or the Brethren for encouragement, the building up of their faith, and to increase their hope in Jesus Christ. In sharing these three important areas of faith-building, many prophetic messages and promises are included in these Epistles of the Bible. As these Epistles are presented, a background of each of the apostles is given. Therefore, in Section Five, the Epistles of Peter, Paul, and John are presented in light of the prophetic teachings regarding 'End Time' coming signs/events.

~⸎~

Chapter 18

THE EPISTLES OF PETER PREDICT UNUSUAL 'END TIME' SIGNS

The Apostle Simon Peter

Simon Peter, as he was named, along with his brother, Andrew, had a fishing business. They apparently conducted this business mostly at the Sea of Galilee. One day Jesus was walking by the Sea of Galilee and saw the two brothers put a net into the sea. Without much introduction to the two men, Jesus told them to come and follow him so that they could fish for people. Interestingly, at once they left their nets and followed him. (Matthew 4:18– 20)

Peter and Andrew followed Jesus Christ in his ministry on earth. Peter was the first of the disciples, and it seems that he never let that position leave his mind. He was the one most often with Jesus Christ, even when the others were not there. There are occasions in the Bible when Peter, James, and John appeared to be leading disciples of Jesus Christ, but Peter emerged as the leader among Jesus's disciples.

Peter was the one who answered Jesus's question about his identity. Peter told him that he was the Messiah, the Son of the living God. Peter even told Jesus he would die for him. But Peter denied Jesus three times during Jesus' trial, as Jesus had predicted. So, after Jesus' crucifixion, His resurrection and ascension, Peter and some of the disciples returned to their former trade of fishing.

One night as they were fishing, Jesus appeared to them, walking on the water. Next, He invited them to breakfast on the shore. At this breakfast, Jesus challenged Peter to the test of 'Love.' Take note of the conversation in John 21:15–21:

1. "Simon, son of Jonas, *lovest* (*agapas*: divine *love*) for me more than fish, bread, and other things?"

2. "Yes, Lord; you know that I *love* (*Philo*: *like or have affection*) for you.
3. "Simon, son of Jonas, do you *love* (*agapas* - divine love) for me?"
4. "Yes, Lord; you know that I *love* (*Philo*: *like or have affection*) you.
5. "Simon, son of Jonas, do you *love* (*Philo*: *like or have affection*) for me? "Peter was grieved because Jesus said unto him the third time if he liked him.
6. "And he said unto him, Lord, you know all things; you know if I love you or like you because you are the all-knowing one. You are God, and you know all things.

Jesus asked Peter three times if he loved him. A crucial moment here— two times Jesus asked Peter if he had divine love for him. Peter refused to answer that he, Peter, had this *Divine love* for Jesus. Rather, Peter had affection for Jesus, he liked Him very much. The third time Jesus did not ask Peter if he had a *Divine love* for him; rather, he asked him if he had an affection or liking for him. Peter was perplexed by this time and told Jesus, He is the Son of God and had all the knowledge (ginoskeis/ gnosis Gk.) of humankind and knew if even a liking for him was true or false. As far as Peter was concerned, deep down within himself, he did not know if he even liked Jesus. But Jesus told him to feed His sheep.

Peter had a big job to do. *First,* he had to feed the lambs, the little babies in Jesus, and help them to grow and become sheep. *Second,* it was Peter's duty to feed the sheep in a twofold manner that they would grow to become greater servants of God. Jesus told Peter of the life he presently had and the kind of life he would have when he became an old man.

Nevertheless, after all the denial, going back to fishing, which was his old way of life, Jesus still told Peter to follow him by imitating his life (Greek - Akolouthei moi—English - imitating me). In other words, Jesus wanted Peter to be like him.

On the day of Pentecost, after the outpouring of the Holy Ghost, the disciples began speaking in other languages. Peter was the one who preached the gospel through the grace of God. Thus, three thousand people became followers of the work of Jesus Christ. Later, Peter received many prophecies about the last days or *End Time signs* that will take place in the sky, on the lands, in the waters, and with the weather.

The Apostle Peter Predicts Unusual Signs in the Sky

The apostle Peter recorded *'End Times'* signs/events as they were revealed to him by the Holy Spirit. Yes, he was present with Jesus Christ throughout all of Christ's ministry on the earth. Peter's epistles are found in the New Testament and are two – 1st & 2nd Peter. In these two Epistles Peter received prophetic information about the *'End Times'*. For example, he writes about "Strange and Unusual Signs/Events in the Sky and Its Great Effect on the Earth." Let us investigate this prophecy and identify the strange eventful signs.

The Strange and Unusual Signs/Events in the Sky and Its Great Effect on the Earth

The apostle Peter states in his *Second Epistle*:

> But the day of the Lord will come as a thief in the night; in which, the heavens shall pass away with a great noise, and the elements shall melt with fervent heat, the earth also and the works that are therein shall be burned up. Seeing then that all these things shall be dissolved, what manner of persons ought ye to be in all holy conversation and godliness, Looking for and hasting unto the coming of the day of God, wherein the heavens being on fire shall be dissolved, and the elements shall melt with fervent heat? (2 Peter 3:10–12)

First, like many of the Old Testament prophets, and the other apostles, the apostle Peter prophesied that the day of God is coming unawares. The prophecy refers to the coming of the Lord as if like a thief in the night. It predicts Jesus Christ's coming. The coming does not indicate the time but a description of what will take place when this day comes. For example, there is going to be a great noise in the sky. In thinking about the noise in the sky, we need to think about the entire universe. The universe will be shaken, and there will be a rearrangement of the planets and stars. God is rearranging and making over some structures. The noise speaks to the movement of all these planets. The earth is only a small part of what is happening here. This noise is so high that it will affect the very building blocks, or atoms, that make up the present heavens and earth. They will melt with very great heat.

Second, the apostle Peter's prophecy predicts that this melting away and rearrangement of the atoms of the universe will then allow the new heaven and new earth to be made by God. He did it once before (Genesis 1; John 1) and he can do it again. Remember that just as the apostle John presented his prophecies, what the apostle Peter predicted will become apparent. The universe will be rearranged or restructured.

For example, a planet will come close to the new earth, which is called the New Jerusalem. All the believers of Jesus Christ will live here. Then, there will be the establishment of a new earth. This new earth will be positioned just outside New Jerusalem. People will move from one to the other by some unknown mechanism. It is not clear if this mechanism for travel and movement will be 'wings' or some 'mind power'. However, all beings will be able to go and visit.

Third, Peter noted that the setting up of this New Jerusalem and the fact that the old heaven and earth will be burned, to have such knowledge of what is going to take place should cause people to live a good and godly life and have real conversations about it. The fact that the Bible says the earth is in danger of burning up one day, would global warming/climate change have anything to do with it?

The first indication that this world system and the entire universe is about to be liquidated or rearranged, Peter warns that the message is from God. It is a prophecy that must be fulfilled. Peter indicated that much of the world is not looking for such a day. However, Peter believed God and looked for this day of the ushering in of the new heavens and new earth.

The Apostle Peter Predicts Unusual Signs on the Earth

Peter predicted that (1) Unusual Event of the Melting Away and Cleansing of the Earth by Fire, and (2), Unusual Criticism of Scoffers and Educators about the Word of God. Note the discussion.

First, Peter predicted the strange and unusual signs/events of the disappearing of the heavens and the *melting away* and *cleansing* of the earth by fire. Therefore, the apostle warns that people should not say that the Lord is slow, nor has he fallen asleep. People need to understand that because of this coming event, God prefers to wait so that people may come to repentance before he destroys the earth with fire.

Second, according to the teachings of the Bible, on this day of God, (1) Israel will be judged, (2) the Gentile nations will also be judged, (3) Jesus Christ the Messiah will establish the thousand-year Kingdom on earth, and (4), Satan will be imprisoned in the abyss (the bottomless pit) for one thousand years.

Third, after 1000 years, Satan will be released from the Bottomless Pit. He caused the nations who lived under the rule of Jesus Christ for the last 1000 years to fight against Jesus Christ, the saints, and the millennial kingdom on earth. At the end of the war against Jesus Christ and his kingdom, God will cast Satan into the lake of fire, the nations will be destroyed, and the old heavens and earth will pass away, as the apostle Peter predicts.

Fourth, following the above-mentioned event, the *'Great White Throne Judgment'* will occur. After this judgment concludes, there will be the appearance of the new heaven and new earth. To have a new heaven and new earth, with the former passed away, means that Peter's prediction has to do with the way God will bring an end to all things when Jesus Christ comes back to the earth.

Fifth, Peter being an Israeli person and a former Jew by religion, looked forward to this new beginning of heaven and earth. Remember—(1) Peter's ministry had much to do with the Jewish believers. (2) In some cases, he had a problem with Gentile believers. He and Paul had words for such an occasion. (3) In Peter's vision on the housetop, he was told to rise, kill, and eat, and his answer was that he had never eaten anything that was unclean.

Sixth, Peter was looking for the destruction of the old heaven and earth and the coming new heaven and new earth. Note what Peter reveals:

> Looking for and hasting unto the coming of the day of God, wherein the heavens being on fire shall be dissolved, and the elements shall melt with fervent heat? Nevertheless we, according to his promise, look for new heavens and a new earth, wherein dwelleth righteousness. Wherefore, beloved, seeing that ye look for such things, be diligent that ye may be found of him in peace, without spot, and blameless. (read 2 Peter 3:8–14)

This prophecy has a reference to the day of God. Many of the Old Testament prophets also prophesied about this day and the meaning that it has for the people of the world and the Jewish nation. The apostle Peter reminds us that this earth we occupy will one day undergo a transformation by "fervent heat from a fire." The time of the event will not be made known to humankind. It will be like the strategy of a thief breaking into someone's house.

The day of God will influence the heavens, the earth, and everything in between. When it happens, the heavens will disappear with a roar, and the elements, building blocks, or atoms that hold together the earth will be dissolved in a fire; everything of the earth will be burned.

It appears there will be a makeover of the earth by the rearranging of atoms. The prophecy warns people to repent and be ready because when God's promise takes place, in this new heaven and new earth only righteousness will dwell. Also, believers should expect such hope to come to pass. The prophecies of the apostle John agree with Peter's prophecies, which are explained in greater detail as they were given to him on the island of Patmos, while he was imprisoned there.

These prophecies should cause believers to live the way the apostle predicted. Also, it should cause people who are not believers to turn to Jesus Christ and be ready to meet him. However, the predictions say that people will doubt the truth that is predicted and believe that God is either dead or has fallen asleep. Note Peter's warning,

> But, beloved, be not ignorant of this one thing, that one day is with the Lord as a thousand years and a thousand years as one day. The Lord is not slack concerning his promise, as some men count slackness; but is longsuffering to us-ward, not willing that any should perish, but that all should come to repentance. But the

> day of the Lord will come as a thief in the night; in which, the heavens shall pass away with a great noise, and the elements shall melt with fervent heat, the earth also and the works that are therein shall be burned up. Seeing then that all these things shall be dissolved, what manner of persons ought ye to be in all holy conversation and godliness. (read 2 Peter 3:8–14 for the full account)

In Peter's warning, he mentioned that scoffers, critics, and educators will come and will criticize the Holy Word of God. This act by such people will be a sign of the 'End Time.'

The Predictions about Scoffers, Critics, and Educators Criticizing the Holy Word of God

The explanation of this sign means that in the last-days critics will be asking, "Where is the promise of his coming?" They will not even wait for an answer; they will then give the answer – **"For since the fathers fell asleep, all things continue as they were from the beginning of the creation."**

It is amazing how the logic and thinking of the world are contrary to the Word of God. But then, could it be that the Believers of Jesus Christ caused people to think this way? People may see the professed believers of Christ living no different from theirs. Therefore, they may get the feeling that God has fallen asleep, and the "followers of the Way" are no more than a joke.

The Apostle Peter Reminds the Believers to be Alert and Sober minded in the 'End Time'

In the last chapter of Peter's First Epistle, he reminds the Believers of Jesus Christ to be alert and sober-minded. Also, they must remember that Satan is behind his plan to turn people away from the truth of God. Because these *'End Times'* signs/events are already occurring, Peter warns the Believers of Jesus Christ to with this scriptural text:

> Be alert and of a sober mind. Your enemy the devil prowls

> around like a roaring lion looking for someone to devour. Resist him, standing firm in the faith, because you know that the family of believers throughout the world is undergoing the same kind of sufferings. (1 Peter 5:8)

The following chapter of Section Five agrees with the prophecies of the apostle Peter. However, Paul's predictions are for encouraging the Believers of Jesus Christ and sharing the hope for them as the End Times are ushered in. Can you imagine changing from a human body into a body that is incorruptible, immortal, holy, and eternal? Let's check it out!

Chapter 19

THE EPISTLES OF PAUL PREDICT UNUSUAL 'END TIME' SIGNS

The Apostle Paul

First, it must be understood that Saul was the persecutor of the followers of the Way. These followers were believers who had accepted the message of Jesus Christ and had committed their lives to him. The Jewish people labeled the followers of Jesus as heretics, and they tried to destroy them. Among those who persecuted the followers of Jesus Christ was a man named Saul. Saul supervised the stoning death of Stephen, one of the leading deacons of the early church. After Stephen's death, Saul was on his way to Damascus to find and prosecute any believer he could find.

In Acts 9:1–17 and 22:3-21, Paul told the story of his life before and after he met Jesus Christ on the road to Damascus. Paul stated that he is a Jew who was born in Tarsus, a city in Cilicia. He studied in the city of Cilicia, under the teachings of the famous teacher Gamaliel. Gamaliel taught him the perfect laws of the Israeli fathers, so he became zealous toward God. Saul, believing that Jesus Christ was a heretic and so were His followers, with his zeal for the Jewish religion, he persecuted, delivered to prison and even death, men and women who followed Jesus Christ.

Paul called to witness the high priest and all the elders who gave him letters to go to Damascus, to bring the Believers of Jesus Christ in bondage to Jerusalem to be punished. However, he testified that on his way to Damascus, the light of Jesus Christ shone upon him, and Jesus spoke to him. Jesus Christ questioned him. He called him by name, Paul. Jesus asked him why he was persecuting Him, Jesus. Paul said he recognized the voice and called out to Him as Lord, and asked him what He wanted him, Paul, to do. Jesus told him to go to the city of Damascus and Ananias, a Believer will cause him, Paul, to receive his sight. See, Paul became blind when Jesus Christ visited him that day he encountered Jesus on the Damascus' road.

Paul stated that on meeting with Ananias of Damascus, he was prayed for, received his sight, was baptized, and received his commission to go and preach to the Gentiles. (see the full story in Acts 9:1–17 and 22:3–21)

Second, the apostle Paul was not in the company of the other disciples in hearing the teaching of Jesus Christ while he was performing his earthly ministry. Paul was not present with the disciples during the ascension event where they saw Jesus Christ leaving the earth and disappearing in a cloud. He did not hear the two angels announce that "this same Jesus that is taken up from you will return in like manner." However, he encountered the resurrected Jesus Christ on his way to Damascus and was commissioned to be an apostle to the Gentiles. While he was praying in the temple at Jerusalem, he fell into a trance, and he saw Jesus Christ, who told him to leave Jerusalem quickly.

In the book of Galatians 1:11–21, Paul testified to the truth of his earthly ministry. He stated that his preaching is not because of men but it was revealed to him by Jesus Christ. The knowledge of Jesus Christ, God, and the things to come, he received from God and Jesus Christ while he was in the Arabian desert. Therefore, with his training about God and Jesus Christ by the Holy Spirit in the Arabian desert, he began preaching the Gospel of Jesus Christ to the Gentile nations. Although Paul's ministry was to the Gentile nations, salvation's message is to all people. Note, **"Even the righteousness of God which is by the faith of Jesus Christ unto all and upon all of them that believe: for there is no difference"** (Romans 3:22-23).

The apostle Paul was a resolute apostle of the Lord Jesus Christ. God used this apostle to write more of the New Testament books than any of the other apostles. The apostle Paul was not with Jesus Christ during His earthly ministry. However, he met Jesus Christ on the road to Damascus while en route seeking to persecute the believers of Jesus Christ. Afterward, God, through the inspiration of the Holy Spirit, revealed to Paul many mysteries that would take place with the believers of Jesus Christ in the *'End Times'*.

The apostle Paul recorded many predictions about the *'End*

Times'. His messages from God were on, (1) the living styles of people during the *'End Times'*, (2) deceivers who will come with a form of godliness but will outright deny the power of God, (3), strange and unusual events or signs that would take place in the sky, on land, in the waters, and with the weather, and (4), one of the greatest mysteries that the apostle Paul received from God and predicted was the removal of believers from the earth to meet the Lord in the sky. Let us now investigate Paul's predictions about unusual events/signs in the sky.

The Apostle Paul Predicts Unusual Signs in the Sky

According to the apostle Paul, some of the strange and unusual events/signs that will take place in the sky in the *'End Times'* will be (1) The Snatching Away of the Believers of Jesus Christ in the 'Cloud' to the Sky, and (2), the Believers of Jesus Christ will receive awards during an Awards Ceremony in Heaven. The chapter discussed each of these events.

First Prophecy - The Snatching Away of the Believers of Jesus Christ in the 'Cloud' to the ***Sky*** (1 Thessalonians 4:13–17)

Many of the prophecies that Paul received from the Lord refer to the sky. He prophesied that Jesus Christ is coming back to take the called-out assembly (church) to the clouds in the heavens. These believers are not members of a church but are all over the world, sometimes called "the church invisible." Because of their faith in Jesus Christ, He will take them to **the sky**, out of this world, before the great tribulation starts on the earth.

First, in 1 Thessalonians 4:13–17, Paul encouraged the believers of Jesus Christ about a future event which today we call "The Rapture" (read the account). However, from the account of 1 Thessalonians 4, Paul predicts that Jesus Christ will come on the clouds in the sky to receive His body of believers.

Second, apparently, the believers at the church of Thessalonica had a problem regarding the believers who had died and wondered what would become of them. Paul, under the inspiration of the Holy Ghost, prophesied that believers, whether dead or sleeping in Jesus Christ, or those still living on earth, Jesus Christ will change them, and they will meet the Lord in the air.

Third, in this prophecy, Paul explained how the Lord himself is coming from heaven with a shout, the voice of the archangel, and with the trumpet of God. The shout, the voice, and the sound of the trumpet of God will awake all the dead believers from the grave first. Then the believers that are still alive will suddenly the Holy Spirit's power will change them and together they will go up to meet the Lord in the air. No doubt this will be a very unusual event that will take place sometime soon.

*Second Prophecy - The Snatching Away of the Believers of Jesus Christ in the 'Cloud' to the **Sky*** (1 Corinthians 15:19–58)

The apostle Paul made another prediction to the believers in the church in Corinth and to all believers who are a part of the "Called-Out Assembly." His prediction concerned believers who die in Jesus Christ and have hope of eternal life. The prophecy is in 1 Corinthians 15:19–58.

First, in the prophecy, it explains that hope only in the present life is a very miserable way to live. Because all individuals since Adam died, people who believe in Jesus Christ will live. The reason people in Christ have a real- life, Jesus Christ is the *'first fruits of the resurrection'*. Therefore, individuals who die in the Lord will one day receive a new body that is immortal and incorruptible (read 1 Corinthians 15:19–58 for the full story)

Second, Jesus Christ promised this new body to the believers on that day when He takes them from the earth to meet Him in the sky. At that time, God will bring all the believers who died in Christ from heaven with Him. At which time, a unification of soul and new bodies will take place in the sky. (see 2 Corinthians 5:1-5)

*Third Prophecy - The New Bodies for the Believers of Jesus Christ are in Heaven, in the **Sky already*** (2 Corinthians 5:1–5)

Jesus Christ promises the new body to the believers. He will give it to them on that day when He resurrects and transforms them from the graves and the earth, to meet Him in the *sky.* The apostle prophesied that whenever the believers become absent from their present bodies, new bodies await them in the heavens. Note what the prophecy guarantees for the Believers in Jesus Christ.

First, it is a fact that when a Believer dies and leaves the house, tabernacle, or physical body, that person already has another body that Christ has made without human functions. This body is an everlasting body and is in heaven awaiting the Believers on their departure from the old earthly, physical, corruptible, mortal body.

Second, there is no comparison of the new heavenly, immortal, incorruptible body with the old earthly, physical, corruptible, mortal body. In this present earthly body, people have pains and physical problems all the time. But in this body, there will be no more pain or sorrow. In fact, the apostle John added to what Paul's prophecy is predicting. It says in Revelation 22:4, that **"God will wipe away all tears from the eyes of those who die in the Lord. Also, there will be no more death, sorrow, crying, and pain because all the former problems of earthly life will pass away."**

Third, the proof of this prophecy is in the scripture which predicts that God has given to the Believers, **"the earnest or down payment of the Spirit."** The Spirit of God who lives in the Believers of Jesus Christ is God's deposit or down payment that the Believers' houses or bodies in heaven are there because of "The Ratified Deed – The down payment of the Spirit" (read 2 Corinthians 5:1–5).

Fourth Prophecy in the Sky – The Apostle Paul's Prediction of the Believers of Jesus Christ's Awards Ceremony (1 Corinthians 3:11–15)

Another unusual event predicted to take place in *the sky* is a judgment called the Bema seat judgment or the judgment seat of Christ. This judgment is an event where the believers will receive rewards for their work on earth, but some will not receive any rewards if they did not do much for the work of Jesus Christ. However, He will not reject these believers from heaven. The Holy Bible states that He will save them "as by fire." They will be safe and saved from an eternal judgment of fire because they received the Lord Jesus Christ as their Lord and Savior.

First, the apostle Paul made the assurance of this judgment clear: "For we must all appear before the judgment seat of Christ; that everyone may receive the things done in his body, per that he hath done, whether it be good or bad" (2 Corinthians 5:10, emphasis added). 'All' means all believers, whether or not they worked for Christ in the Spirit or the flesh. But those who worked in the flesh will NOT have crowns to cast at the feet of Jesus Christ during the great celebration in high praise, honoring Jesus Christ, the Lamb and the Lion.

Second, this event will take place immediately after Jesus Christ takes His Believers to meet Him in the sky. In another prophecy, Paul explains this Judgment Seat of Christ and the giving out of rewards that the Bible calls "crowns" in the prophecy. At the Bema Seat, Jesus Christ will give out "Crowns" to the Believers who have worked for Him while on the earth. Remember, this crown ceremony has nothing to do with Salvation. Salvation is a gift which God gives to any person who asks Jesus Christ to forgive them of their sins and come into their lives. Also, no person has to work for this salvation. However, the Crowns that Jesus Christ will give out at the Bema Seat Judgment, are on the Believers' works which they did in the Spirit of Christ after they became Believers of Jesus Christ.

Third, 1 Corinthians 3:11–15 explains the test for the Crowns this way. The Believers of Jesus Christ have the option to build upon the foundation of Jesus Christ, gold, silver, precious stones, wood, hay, and stubble. Each of these represents works of the Believers, either in the person's flesh doing the work or in the spirit of God. However, every person's works the Judge at the Bema Seat (Jesus Christ) will try by the testing fire of God. God's testing fire will easily burn the works that resemble wood, hay and stubble or trash. The testing fire of God will prove that the Believer did such works in the flesh and not in the Holy Spirit. Therefore, these believers will not receive a reward. But the works of gold, silver, and precious stones, they will pass the test because fire makes these items better. Such believers will receive an *award of a Crown* because they did their works in the power of the Holy Spirit and not in their flesh for self-glory. However, they will not lose their salvation. This judgment has nothing to do with a person's salvation. It is for rewards or crowns.

The Apostle Paul Predicts Unusual Signs on Land

The Apostle Paul's Prophecies of Unusual Events on Land include (1) The Unusual Events or Signs of Perilous Times throughout the World in the *'End Times'*, (2) The Unusual Events/Signs of Great Deception Targeting the Believers of Jesus Christ throughout the World in the *'End Times'*, and (3), The Unusual Events or Signs of Spiritualism with Seducing Spirits and Doctrines of Devils. Again, the chapter presents and discusses each topic.

Prophecy No. 1: The Unusual Events or Signs of Perilous Times throughout the World in the 'End Times' (2 Timothy 3:1-5)

Under the inspiration of the Holy Spirit, Paul had a message that somewhat relates to the message that Jesus Christ shared with his disciples regarding signs of the end times. Paul prophesied about significant activities on land. In 2 Timothy 3:1-5, the apostle predicts the character of people in the last days or *End Times*.

Paul states that people must be aware, in the last days' or end times perilous times shall come. The signs of 'End Time' will be Perilous Times. Note the conditions:

- People will be self-lovers. This last-days' condition will manifest itself in the love of money, boasting about self, and showing lots of pride.
- People will be without love for others.
- They will become abusive to each other.
- Children will become disobedient to their parents.
- Overall, individuals will become ungrateful, unforgiving, slanderous, without self-control when conducting business or having a conversation, and brutal.
- They will hate people who are trying to be good and will be very conceited in behavior and attitudes.
- They will show that their lives are unholy.

- They will have a form of godliness, but no power of God will be present in their lives.
- Their love will be toward pleasure, just having fun.

This prophecy agrees with the other apostles' prophecies that predicted that people's behavior in the *'End Times'* would be like those in the time of Lot and Noah. They too had no time for God, only pleasure. The Believers of Jesus Christ are to reject this kind of life.

Prophecy No. 2: The Unusual Events or Signs of Deception (2 Timothy 3:6-7)

Another of the signs of the *'End Times'* will be a display of deception. In 2 Timothy 3:6–7, Paul predicts that much lust among people will be on display. Men will take advantage of women and deceive them. Deception will not spare the Believers of Jesus Christ. Deceivers will "worm" or twist their way into peoples' homes. They will target the women then try to gain power over them and lead them into all kinds of evil desires.

It is interesting to note here, that these deceivers are "always learning but never able to come to a knowledge of the truth." These are the "experts"; they will claim to know everything. People will believe in their deception because they are on TV and seen in the news. These people, however, will have no clue of the truth of the prophetic Word of God. They will reject prophecies about the rapture of the church, the great tribulation, and the Second Coming of Jesus Christ. Their minds will be numb to the truths of God, yet they will deceive many people.

Another prediction about the great deception that will come upon the world will be a significant influence on the called-out assembly of Jesus Christ. The prophecy warns the believers that they should "Let no man deceive you by any means: for that day, shall not come, except there comes a falling away first, and that man of sin or the son of perdition, the Holy Spirit will allow him to reveal himself" (2 Thessalonians 2:3).

The apostle Paul predicts signs on "that day." *First,* it refers to the falling away of many who professed to know Jesus Christ. They will turn to the 'deception of the *'End Times'*. *Second,* "That day" predicts the apostasy that was coming upon the church. Some people will accept the teaching that there is a vicar (a mere sinful man) who stands between God and humans and can forgive the sins of humankind. The teaching here will be an attempt to rob the position of Jesus Christ, who is the only mediator between God and humankind. Note what the scripture states in 1 Timothy 2:5—"For there is one God, and one mediator between God and men, the man Christ Jesus" (emphasis added). Apostasy is when any church or people strip Jesus Christ from His position of being the mediator between God and humans and try to place sinful, depraved human beings on the level of Jesus Christ.

These things will come to pass as *"the man of sin* or *the Antichrist"* comes to the scenes in the *'End Times'*. The falling away of believers will continue until the rapture of the called-out assembly. However, the activities of the Antichrist will continue throughout the great tribulation and until the Second Coming of Jesus Christ to the earth.

Prophecy No. 3: The Unusual Events or Signs of Spiritualism with Seducing Spirits and Doctrines of Devils

Paul also prophesied that at the end times there will be a rise in spiritualism. Note the prediction:

> Now the Spirit speaks expressly, that in the latter times some shall depart from the faith, giving heed to seducing spirits, and doctrines of devils; Speaking lies in hypocrisy; having their conscience seared with a hot iron. (1 Timothy 4:1–2)

Watch out for liars who come to you and change the scriptures and teachings of the Bible to which you held for so long. Remember the apostle Paul prophesied that these things would come to pass, and it will be in the *'End Times'*. The prophecy states that the Spirit of God says that some will abandon the faith and follow deceiving spirits and things taught by demons.

Many of the TV programs today producers based them on deceiving spirits and things taught by demons. Therefore, they produce programs, teachings, and activities that are demonic. The same programs which are works of deception have gone into the church of Jesus Christ and have its effects on the Believers of Jesus Christ and *Church Goers*.

Many church followers take part in such activities. Some of the activities have moved into the church. Note, one of the modern activities present in many places of worship is the program "Praise and Worship." There is nothing wrong with praising God and lifting Him up and giving Him all the glory and praise. But whenever any program in a place of worship takes over, and it leaves the Word of God out, pastors and teachers will not have the time to teach and lift up the Believers. Therefore, with only the feeding of emotions only, and no grounding in the Word of God, Satan uses the opportunity to lead *church followers* to spiritualism, a doctrine of demons that the apostle Paul predicts to take place in the "End Time."

We have seen a continuity of the epistles of the apostles Peter and Paul about the End Times Signs. The third chapter of Section Five is about the prophetic signs that the apostle John received from the glorified Christ while on the Island of Patmos in exile. There are so many strange and unusual events that readers will witness here. Let's get to it!

~✣~

Chapter 20

THE EPISTLES OF JOHN PREDICT UNUSUAL 'END TIME' SIGNS

The Apostle John

The apostle John, known as "the beloved," received many prophecies from Jesus Chait regarding the *'End Times'* signs/events in the sky, on land, in the waters, and with the weather. Some of the prophecies he received while he was with Jesus Christ as one of His disciples. He, like Matthew, Luke, and John, were present during the teachings of Jesus Christ. But while he was in exile on the Island of Patmos, Jesus Christ appeared to John in His glorified form and communicated to him prophetic signs of the 'End Time.' The presentation of the events or signs John saw about the future was very strange and unusual. Some of the signs were so unusual, John asked the angel who was escorting him in this spiritual revelation, to explain the meanings.

In this portion of the chapter, the prophetic 'End Time' signs/events that the apostle John saw occurring in the sky, on the lands of the earth, the waters and seas, and the weather, the chapter presents and discusses. These are, (1) the prophetic events/signs to occur in the sky, (2) the prophetic events/signs to occur on the land, (3) the prophetic signs or events to occur in the waters and seas, and (4), those prophetic eventful signs that will occur with the weather. Let us investigate these prophetic predictions of the apostle John.

The Apostle John Predicts 'End Time' Signs in the Sky

Readers must understand that the prophetic signs/events that the angel showed the apostle John is the entire Book of Revelation. Therefore, the purpose of this book is not to rewrite the Book of Revelation. Some prophetic 'End Time' signs/events are in greater detail than others. With this said, let us now investigate the many strange and unusual signs/events that will occur in the sky. So, the first presentation is about signs/events of the sky. These are (1) the Sign of the Believers' Prepared World. This section John received before his Patmos experience. (2) the signs or events Believers must Know and prepare themselves to face them. This second portion is about the Patmos experience of John.

The Apostle John Predicts that Jesus Christ Is Preparing a World for His Believers in Heaven and He Will Return

Jesus Christ was aware of His time on earth. Therefore, one day he explained to His disciples that He will leave them. During the discussion, He told them that He had to go away, and He will come back for them. However, Jesus Christ was very clear with his disciples about this place He will prepare for them and all other believers. He explained to them (1) where he was going, (2) why his going was important, (3), what promises He will fulfill while in the heavens, and (4), they must worry about nothing because He has the power to prepare this place, for, He and the Father are One. Therefore, being God, He can prepare the promised-world and will bring it back toward the earth, through the sky, for them to live in forever. Jesus Christ is coming back again.

Readers may read this prophecy in John 14:1-12. However, read this first part of the promise:

> Let not your heart be troubled: ye believe in God, believe also in me. In my Father's house there are many mansions: if it were not so, I would have told you. I am going to prepare one for you. And if I go and prepare a place for you, I will come again, and receive you unto myself; that where I am, there ye may be also. (see John 14:1–12)

First, the above prophecy gives hope to any person who is following Jesus Christ and knows Him as Lord and Savior. The hope is, regardless of the various happenings that are taking place in the world today, the followers of Jesus Christ have the hope of seeing Him someday coming to get them.

Second, we must understand, before Jesus Christ fulfills the above prophecy, there are other signs that must first take place. 1. There is a need for the Believers of Jesus Christ to prepare to meet the Lord in the sky, and 2, people need to know, that those who are part of the Called-Out-Assembly of Believers will not only go to heaven, but each believer can receive *seven awards* when they meet at the Bema Seat Judgment in the sky. The apostle John received this information while in exile on the Island of Patmos. Therefore, the Believers of Jesus Christ should know, and they should prepare themselves and acknowledge the signs and those to come. Also, people who will become Believers will also have the same opportunity to enter the place Jesus Christ is preparing.

The Apostle John Predicts Seven Award Jesus Christ Will Give to His Believers Who Overcome Trials on Earth

While John was in exile on the Island of Patmos, on the Lord's day while in worship, Jesus Christ appeared to him and revealed to John information about His Believers on the earth. Because the time is short and Jesus is coming back for them soon, He gave John seven letters, each designated for each of the seven churches. In each letter, the Lord expressed the *strengths*, *weaknesses*, *opportunities*, and *threats* that He saw in the lives of His Believers. Jesus knowing that they will endure the *threat* of many tribulations gave each believer an *opportunity* to overcome their *weaknesses*, and to receive rewards/crowns in heaven (Revelation 1-3). Therefore, in each letter, as recorded by John in the Book of Revelation, these 'promises' and 'blessings' the apostle John presents. So, in the sky/heaven, the Believers of Jesus Christ who are victorious over trials will receive awards.

We must note here, these awards that Jesus promises to the believers in the seven churches are not *the Crowns* that Jesus Christ will give out at the Bema Seat Judgment. The Awards of Crowns are different. The Crown Awards are for building on the foundation of the work of the Church – *for Workers*. These Awards of Crowns are in great detail in the author's book on Amazon, "THE MYSTERY: Humans Becoming Immortals." However, let us now study these *Overcomers Awards*.

Overcomers Award No. 1: Revelation 2:7 states that the followers from the called-out assembly of the *Ephesus Church Group* who are victorious in overcoming trials, Jesus will give that person the right to eat from the tree of life, which is in the paradise of God.

Overcomers Award No. 2: Revelation 2:11 states that the victorious believers of the called-out assembly of the *Smyrna Church Group*, "The Second Death" will not hurt. This "Second Death" refers to the eternal separation from God in a place the Bible names, *the Lake that burns withfire and brimstone.*

Overcomers Award No. 3: Revelation 2:19 states that the victorious believers of the called-out assembly of the *Pergamum Church Group* will (1) receive some of the hidden manna, (2) receive a white stone with that person's new name written on it, and (3) only that person will know the new name.

Overcomers Award No. 4: Revelation 2:26-29 states that each victorious believer who carries out the will of Jesus Christ to the end, such believer of the called-out assembly of the *Thyatira Church Group* will receive (1) authority to rule over the nations, and (2), the *morning star.*

Overcomers Award No. 5: Revelation 3:5 states that the followers from the called-out assembly of the *Sardis Church Group* who are victorious in overcoming trials, Jesus will (1) have that person dressed in white clothes. (2) He will never have his or her name blotted out of the book of life, and (3) He will acknowledge that person before the Father and His angels.

Overcomers Award No. 6: Revelation 3:12 states that the followers from the called-out assembly of the *Philadelphia Church Group* who are victorious in overcoming trials, Jesus will (1) set that person as a pillar in the temple of God permanently, (2) He will write on that person, the Name of His God and the name of the city of His God, the New Jerusalem, which is coming down out of heaven from God, and (3), He will also write on such person His new name.

Overcomers Award No. 7: Revelation 3:5 states that the followers from the called-out assembly of the *Laodicea Church Group* who are victorious in overcoming trials, Jesus will give him or her the right to sit with Him on His throne, just as he sits on His Father's throne.

These awards are to the "Believers of the Church of Jesus Christ." Each church mentioned is a "Period" that Christ Called-Out-Assembly of Believers will undergo throughout history. Therefore, the opportunity to attain these awards is to every believer of Jesus Christ. Therefore, the apostle Peter reminds believers to "live holy and godly lives as they look forward to the day of God and speed its coming" (see 2 Peter 3:11-12).

The Apostle John Predicts Seven Major Signs to Occur in the Sky–Heaven

Jesus Christ revealed to the apostle John many signs/events up in the sky/heaven. Unlike many other prophets and apostles, the Spirit of God took the apostle John to heaven to view the 'End Time' eventful signs. While in heaven, John saw many strange and unusual activities. Revelation 4 presents some of the activities. What John saw, Jesus Christ told him to send the information in letters to His seven churches (See Revelation 4:1-11).

Event No. 1: The Visit to the Throne of God in Heaven

In the eleven verses of the fourth chapter of Revelation, the apostle John on arrival to heaven, he saw the Throne of God, and heard many strange and unusual things:

1. He heard a voice like a trumpet that speaks.
2. He saw a throne, and someone was sitting on it. The description of the throne was of jasper and ruby. Around the throne was a shining rainbow of emeralds.
3. Around the throne, there were twenty-four elders dressed in white, with crowns of gold on their heads. They were also sitting on thrones.
4. Flashes of lightning, rumblings, and peals of thunder came from the throne. In front of the throne were seven lamps which were blazing with light. Also, in front of the throne was something that looked like a sea of glass.
5. In the center of the throne, were four living creatures. Eyes covered the creatures' entire bodies. The creatures were strange looking. One creature looked like a lion, the second looked like an ox, the third had a face like a man, and the

fourth looked like an eagle with wings out ready to fly.

6. Each of the four living creatures had six wings. On top and under the wings were eyes. The creatures praise the Lord God Almighty. In their continuous praise, they said, "Holy, holy, holy is the Lord God Almighty, who was, and is, and is to come."
7. As the living creatures gave glory to the Lord God Almighty, the twenty-four elders got off their thrones, fell down before the Lord God Almighty who lives forever and ever, and they worship Him. As they worship Him who lives for ever and ever, they lay their crowns before the throne of God Almighty and say: "You are worthy, our Lord and God, to receive glory and honor and power, for you created all things, and by your will, they were created." (see Revelation 4:1-11)

Event No. 2: The Releasing of the Seven Scrolls of the Great Tribulation Judgment on the Earth

The second vision John saw of strange and unusual eventful signs in the sky. God on His throne held a scroll of judgments. He sought for a worthy person in heaven, on the earth, under the earth to come and take the judgments and execute them on the earth. However, there was no person found worthy to perform the act. However, the Lamb, the Lion of Judah came forth and took the judgment-scroll from the hand of God (Revelation 5:1-14). When the Lion-Lamb received the judgments from the hand of God, three amazing signs or events took place. *First,* thousands and thousands of angels encircled (1) the throne of God Almighty, (2) the living creatures, and (3), the twenty-four elders, and said with a loud voice: **"Worthy is the Lamb who was slain, to receive power and wealth and wisdom and strength and honor and glory and praise!"**

Second, all the creatures of heaven, on the earth, under the earth, and those in the sea were saying or speaking: "To him who sits on the throne and to the Lamb be praise and honor and glory and power, forever and ever!"

Third, the four living creatures said, "Amen," and the twenty-four elders fell down and worshiped God and the Lamb.

The above has many strange and unusual signs or events which will take place in the sky/heaven. Apart from all the worship given by the four creatures, the thousands of angels, the twenty-four elders – the Saints of the New and Old Testament, all the creatures in heaven, on earth, under the earth, and those in the sea will praise God and the Lamb. Can you imagine hearing the birds, fish, and other creatures shouting praises to God Almighty and the Lamb? Another point here, regardless of what people think or believe, one day, and more than once, people will have to bow and fall down before God and acknowledge Him as the Creator and the One who lives forever and ever, or from ages to ages.

Event No. 3: Three Sets of Seven Judgments the Lamb Will Conduct from Heaven During the Great Tribulation on Earth (Revelation 5-19)

The events or signs of the Great Tribulation will take place on the earth. But the Lamb will control the operation. He will conduct it from heaven. In Revelation 5, the scrolls of judgments that the Lamb took from the hand of God, were the 21 judgments He must pour out on the earth. The first seven are the seal judgments which the Lamb will execute. However, the other judgments, the seven Trumpets, and the seven Vials, the Lamb will allow special angels to execute them. The Voice from the sanctuary in heaven controlled the actions of the angels. On the conclusion of the Great Tribulation judgments, heavenly beings conducted a fourfold Hallelujah ceremony. There is great rejoicing in heaven. Also, just before the King of Kings and Lord of Lords, the Righteous Judge, and Messiah leaves heaven for earth, the Marriage Ceremony of the Lamb takes place. Many signs will occur in the sky and on the earth as He leaves heaven for the earth. There on earth, He will take part in the Battle of that Great Day of God Almighty at Armageddon (Revelation 16:16).

Event No. 4: The War in Heaven–Satan Cast Out (Revelation 12:7-9)

To understand this event, some history of Satan is needed.

First, after Satan rebelled against God he was cast out from the Mountain of God. However, he received permission to enter heaven from time to time. Note the proof, **"Now there was a day when the sons of God came to present themselves before the LORD, and Satan also came among them" (Job 1:6).**

Second, Satan today is the accuser of the Brethren (Revelation 12:10). During the Great Tribulation, God will get rid of Satan from heaven altogether as previously presented. His role as the accuser of God's people in heaven will not be necessary because Jesus Christ will have taken the Believers from the earth. They will now be in their new eternal, incorruptible, immortal, and spiritual bodies.

Third, the event that will occur in heaven during the Great Tribulation on the earth will be a war between Satan, his angels, and Michael and his angels. Satan will have to leave heaven, but he will refuse to do so. So, he will declare war. Michael and his angels will defeat Satan and cast him out of heaven, forever. Revelation 12:7-8 of this prophetic passage tells us that, **"...there was war in heaven: Michael and his angels fought against the dragon, and the dragon fought and his angels.**

Fourth, after Michael the Archangel casts Satan out of heaven, Satan will strengthen the Antichrist and the False Prophet to make war against the Israeli people. The nation of Israel will undergo much persecution by the satanic force of Satan, the Antichrist/Beast, and the False Prophet. Many strange and unusual eventful signs will occur during the last three and a half years of the Great Tribulation.

Event No. 5: The Marriage of the Lamb and the Return of the King of Kings and Lord of Lords from the Sky

During the Great Tribulation on the earth, in heaven, just before Jesus Christ leaves heaven, the Marriage of the Lamb which this section previously mentioned, will occur. However, note the Hallelujah Ceremony (see Revelation 19:1-9).

The *first* Hallelujah is given because **"Salvation and glory and power belong to God, for true and just are his judgments. He has condemned the great prostitute who corrupted the earth by her adulteries."** The Lamb ends the false religious system of Babylon. This religious system is in our world today in disguise. However, after the believers of Jesus Christ are taken from the world prior to the Great Tribulation, this system will usher in the Beast/Antichrist. Remember, this religious system is controlled by Satan.

The *second* set of Hallelujahs is given because **"the smoke of the Babylonian Harlot goes up forever and ever."** Also, **"Praise our God, all you, his servants, you who fear him, both great and small!"** Heaven will rejoice over the destruction of the religious system that persecuted and killed many of the Saints of Jesus Christ.

The *third* Hallelujah given– **"For our Lord God Almighty reigns. Let us rejoice and be glad and give him glory! For the wedding of the Lamb has come, and his bride has made herself ready. Fine linen, bright and clean, was given to her to wear."** Note, fine linen stands for the righteous acts of God's holy people. This is an identification that the Bride of the Lamb refers to the Righteous Saints of God who have been in heaven since the Rapture of the Believers. Now, the party will leave heaven and come through the sky with the King of Kings and Lord of Lords, to execute judgments on the earth, and set up His Millennial Kingdom. At which time, the Nation of Israel will turn to God and will receive their long-promised blessings.

Event No. 6: The Coming New Heaven and the New Earth

The apostle John prophesied, **"And I saw a new heaven and a new earth: for the first heaven and the first earth were passed away; and there was no more sea"** (Revelation 21:1). This truth that John is predicting was also predicted by Isaiah, the Old Testament prophets, and the apostle Peter.

The prophet Isaiah prophesied, **"For, behold, I create new heavens and a new earth: and the former shall not be remembered, nor come into mind"** (Isaiah 65:17). Isaiah also prophesied, **"For as the new heavens and the new earth, which I will make, shall remain before me, saith the LORD, so shall your seed and your name remain"** (Isaiah 66:22). Regarding the new earth, Isaiah prophesied:

> And they shall bring all your brethren for an offering unto the LORD out of all nations upon horses, and in chariots, and in

> litters, and upon mules, and upon swift beasts, to my holy mountain Jerusalem, saith the LORD, as the children of Israel bring an offering in a clean vessel into the house of the LORD. (Isaiah 66:20)

The apostle Peter reminds people, **"Nevertheless we, according to his promise, look for new heavens and a new earth, wherein dwelleth righteousness"** (2 Peter 3:13).

Event No. 7: The Holy City—the New Jerusalem That Will Come Down from God (Revelation 21:1)

The apostle John continued in his prophecy:

> And I John saw the holy city, the New Jerusalem, coming down from God out of heaven, prepared as a bride adorned for her husband. And I heard a great voice from heaven saying, Behold, the tabernacle of God is with men, and he will dwell with them, and they shall be his people, and God himself shall be with them, and be their God. And God shall wipe away all tears from their eyes; and there shall be no more death, neither sorrow, nor crying, neither shall there be any more pain: for the former things are passed away. And he that sat upon the throne said, Behold, I make all things new. And he said unto me, Write: for these words are true and faithful. And he said unto me, It is done. I AM The Alpha and Omega, the beginning, and the end. I will give unto him that thirst to drink of the fountain of the water of life freely. (Revelation 1:2-6)

The above prophecy says a New Jerusalem will come down from God out of heaven. This city will be exquisite. Jesus Christ prepared it for his bride (His Church), for He is her husband. This city will represent God, the Lamb, and people living together. Once this takes place, the people who live in this city will have a good life because all things God will make new. In the new world system and in New Jerusalem, God—the Alpha and the Omega, the beginning, and the end—gives the spring water of life without payment **"to the thirsty."** That same spring water of life is Jesus Christ. He told the woman at the well in the Gospel of John that **"whoever drinks from the water that he gives will never thirst**

again." Jesus Christ is the sustainer of everlasting life in the New Jerusalem and in the new world Theocracy, the way God always wants.

The old-world system that Satan contaminated with sin is now over, and God has made all things new. The apostle added that this promise is for all people today, and God wants to give this heritage to all his conquerors who now live on earth. The prophecy states, **"He that overcometh shall inherit all things; and I will be his God, and he shall be my son"** (Revelation 21:8). He seeks to be the source of their thirsty souls, and he wants to be their God. So, whoever will drink the water that Jesus gives will become the children of God. The other reason why God wants to be the source of life for humankind is because of His warning in Revelation 21:8. It declares, "But the fearful, and unbelieving, and the abominable, and murderers, and whoremongers, and sorcerers, and idolaters, and all liars, shall have their part in the lake which burneth with fire and brimstone: which is the second death."

Life in PARADISE (Revelation 21:9–22:5)

The holy city, the New Jerusalem, the angel showed and described to the apostle John by the last of the seven angels. This angel was the one who administered the last seven plagues on the sky, the land, and the waters. He is the one who showed the apostle John the view of Paradise and how God's people will live there.

First, the holy city, the New Jerusalem will come down from God out of heaven. In it, the glory of God will resemble rare jewels, such as jasper and crystal. The city John saw, Jesus Christ made it of gold that is as clear as crystal.

Second, the city is a square with four high walls made of jasper. In each of the four walls are three gates, which makes twelve gates in this city. At each gate, there is an angel who guards it. Therefore, the number of angelic guards for the city is twelve. At each gate to the city is a pearl and on each gate is the name of one of the twelve tribes of the sons of Israel.

Third, Jesus Christ builds the city on twelve foundations. On each foundation, The Lamb inscribed the names of His Twelve Apostles. Also, He adorned the twelve foundations with twelve different jewels. These jewels are jasper, sapphire, agate, emerald, onyx, carnelian, chrysolite, beryl, topaz, chrysoprase, jacinth, and amethyst. The interpretation in this passage implies that the names of Simon Peter, James, the son of Zebedee, John–the son of Zebedee, Andrew–the brother of Simon Peter, Philip of Bethsaida, Thomas– called Didymus, Nathaniel Bartholomew, Matthew–Levi of Capernaum, James–the son of Alphaeus (the Lesser), Simon the Zealot–the Canaanite, Thaddeus or Jude, and Paul–formally called Saul, are the names inscribed in the foundations of the New Jerusalem.

Fourth, the city has streets of pure gold that look like transparent glass. In the city, the temple is the Lord God Almighty and the Lamb. The Lamb is the light of the city so there will be no night there. In His light, the nations to walk in, and the kings of the new earth will bring to New Jerusalem the glory and honor of the nations. The people in the city will also be able to see the face of God and the Lamb. His name will be on their foreheads and the people will reign with him forever and ever because in the city is the throne of God and the Lamb.

Fifth, in the city, is the River of Life, which is as bright as crystal and flows from the throne of God and of the Lamb through the middle of the city. On each side of the river is a tree of life that grows twelve kinds of fruit each month. Also, the leaves of this tree give healing to the nations.

Sixth, the city will have nothing that is unclean and will have no one who is detestable and false. The only people who will be in the city are those that the Lamb wrote their names in His Book of Life.

Besides these unusual events in the sky/heaven, John predicts many horrifying signs that will take place on the earth. Among these *End Time Signs* on the earth will be an event in which some people will be mysteriously taken from the earth. Let us now check it out!

The Apostle John Predicts 'End Time' Signs on the Land–affecting the Sky, Waters, and Weather

Event No. 1: The Rapture of Snatching Away of the Believers of Jesus Christ from the Earth to the Cloud

John predicts many horrifying eventful signs that will take place on the earth. But among these End Time Signs, for the believers, this event will be a good one. But to families and friends who might be left behind, it will be horrifying. The fact is, there will be the "Rapture" or the "Caught Up" of the Believers of Jesus Christ from the earth to the sky. This is seen in the event in Revelation 4:1-2 which states:

> After this I looked, and there before me was a door standing open in heaven. And the voice I had first heard speaking to me like a trumpet said, "Come up here, and I will show you what must take place after this." At once I was in the Spirit, and there before me was a throne in heaven with someone sitting on it.

The prophetic scripture is a demonstration of the "Rapture" of the Church to heaven. As presented by the apostle Paul and the "Snatching Away," John also heard "a vice speaking," "the sound of a trumpet," and the activity of the Holy Spirit during the Rapture of the church of Jesus Christ. Can you imagine the confusion on earth when only people who know Jesus Christ as Savior will be suddenly taken from the earth to heaven? There will be much commotion throughout the world with people looking for missing wives, husbands, children, relatives, friends, and co-workers. Newspapers and social media will have much to communicate about the missing people from the earth. The author's book, THE MYSTERY: Humans Becoming Immortals, explains this event in full detail. Readers may purchase the book on Amazon.

Event No. 2: The Great Tribulation on the Earth

The Great Tribulation will occur on the earth immediately following the Rapture of the Believers of Jesus Christ. The previous section mentioned some details about the Great Tribulation. However, during the Great Tribulation, many strange and unusual eventful signs will take place in the sky, on the land, in the waters, and with the weather. The Great Tribulation events or signs will occur in 21 judgments. Seven are *Seal Judgments* (Revelation 6:1-8:6), seven are *Trumpet Judgments* (Rev. 8:1-15:8), and the last seven are *Vial Judgments* (Revelation 16:1-18).

(a) The Judgment of the Seven Seals (Revelation 6:1-8:6)

The Seven Seals will have eventful signs such as (**1**) *A World Dictator* (Rev. 6:1-2), (**2**) *A Great World War* (Rev. 6:3-4), (**3**) *A World Famine* (Rev. 6:5-6), (**4**) *A Great Slaughter on one-fourth of the World's People* (Rev. 6:7-8), (**5**) *The Killing of Many Tribulation Saints* (Rev. 6:9- 11), (**6**) *Strange Activities with the Weather and the Sealing of the 144,000 Israeli preachers and their accomplishments* (Rev. 7:1-17), and (**7**), *The Greatest Fear the World will Ever Experience* (Rev. 8:1-6).

(b) The Judgment of the Seven Trumpets (Revelation 8:1-15:8)

The Seven Trumpets will have eventful signs such as, (**1**) *The Greatest Storm on Earth* (Revelation 8:7); (**2**) *The Greatest World Disturbance of The Seas* (Revelation 8:8-9); (**3**) *The Greatest Disturbance to The Waters of The World* (Revelation 8:10-11); (**4**) *The Greatest Disturbances in the Skies Affecting the Lighting of The Earth* (Revelation 8:12-13); (**5**) *The Demonic Attack from the Underworld on the People of the Earth as the Abyss is unlocked* (Revelation 9:1-12); (**6**) *The World's Greatest Army* (Revelation 9:13-14), *The Angel with the Little Book of Prophecy* (Revelation Ch. 10:1- 11), and *The Two Prophesying Witnesses* (Revelation Ch. 11:1-14); and (**7**)*The Greatest Kingdom Victory Ever* (Revelation 11:15-15:8).

The Seventh Trumpet Judgment has about *fifteen components of signs and other activities*. At the end of these components, the seven Vial Judgments will start. However, listed are the *Prophetic Components of the Seventh Trumpet Judgment.*

*(c) Prophetic Components of the Seventh Trumpet (*Revelation 11:15-15:8)

During the Seventh Trumpet Judgment, Fifteen Signs will manifest themselves. They are: (**1**) *The Victory of Jesus Christ, The Messiah* (Revelation 11:15-16); (**2**) *The Sign of the Angry Nations* (Revelation 11:18); (**3**) The Sign of the Open Temple (Revelation 11:19); (**4**) *The Sign of The Woman Clothed With The Sun* (Revelation 12:1-2); (**5**) *The Sign of the Great Red Dragon* (Revelation 12:3-6); (**6**) *The Sign of The War In Heaven*(Revelation 12:7-8); (**7**) *The Sign of the Casting of Satan, The Old Serpent, The Devil Out of Heaven* (Revelation

12:9); (**8**) *The Sign Foretelling The Triumph Of God's Kingdom* (Revelation 12:10-12); (**9**) *The Sign of Satan's Persecution Of The Woman* (Revelation 12:13-17); (**10**) *The Sign of The Beast With The Seven Heads And Ten Horns* (Revelation 13:1-10); (**11**) *The Sign of The Beast Coming Up Out Of The Earth* (Revelation 13:11-18); (**12**) *The Sign of The Singing of the New Song Before the Throne to The LAMB* (Revelation 14:1-5); (**13**) *The Sign of The Angel With The Everlasting Gospel* (Revelation 14:6-7); (**14**) *The Sign of The Fallen Evil Powers* (Revelation 14:8-20); (**15**) *The Sign Of Preparation For The Vial Judgments* (Revelation 15:1-8).

All these signs are within these approximately five chapters of Revelation 11:15-15:8. We will see these former systems and organizations ending, which will cause The Greatest Kingdom Victory Ever. Jesus Christ The Messiah is Victorious. Let us look at this victory.

(d) The Judgment of the Seven Vials (Revelation 16:1-18)

The Seven Vial Judgments will have events or signs such as (**1**) *The Great Disease in The Land on The Beast Worshippers* (Revelation 16:2); (**2**) *The Great Attack on The Sea, All Living Creatures in the Sea will Die* (Revelation 16:3); (**3**) *The Greatest Contamination of The Waters on Earth by Blood* (Revelation 16:4-7); (**4**) *The Greatest Effect of Climate Change Around the World* (Revelation 16:8-9); (**5**) *The Greatest Attack of the Plague on The Throne Of the Beast* (Revelation 16:10-11); (**6**) *The Great Battle on the Great Day of God Almighty – Armageddon* (Revelation 16:12-16); and (**7**), *The Great Battle of Almighty God at Armageddon* (Revelation 16:17-21).

Many of the above signs will have a connection to the attacks on the City of Jerusalem, the land of Israel, and the Israeli people, by the Antichrist/Beast, the False Prophet, and the Gentile nations. God will judge the Gentile nations during the battle at Armageddon (Revelation 12:13-17).

Event No. 3: The Return of the King of Kings & Lord of Lords to the Earth

11. And I saw heaven opened, and behold a white horse; and he that sat upon him *was* called Faithful and True, and in righteousness he doth judge and make war.
12. His eyes *were* like a flame of fire, and on his head *were* many crowns; and he had a name written, that no man knew, but he himself.
13. And he *was* clothed with a vesture dipped in blood: and his name is called The Word of God.
14. And the armies *which were* in heaven followed him upon white horses, clothed in fine linen, white and clean.
15. And out of his mouth goeth a sharp sword, that with it he should smite the nations: and he shall rule them with a rod of iron: and he treadeth the winepress of the fierceness and wrath of Almighty God.
16. And he hath on *his* vesture and on his thigh a name written, KING OF KINGS, AND LORD OF LORDS.
17. And I saw an angel standing in the sun; and he cried with a loud voice, saying to all the fowls that fly in the midst of heaven, Come, and gather yourselves together unto the supper of the great God;
18. That ye may eat the flesh of kings, and the flesh of captains, and the flesh of mighty men, and the flesh of horses, and of them that sit on them, and the flesh of all *men, both* free and bond, both small and great.
19. And I saw the beast, and the kings of the earth, and their armies, gathered together to make war against him that sat on the horse, and against his army.
20. And the beast was taken, and with him the false prophet that wrought miracles before him, with which he deceived them that had received the mark of the beast, and them that worshiped his image. These both were cast alive into a lake of fire burning with brimstone.
21. And the remnant were slain with the sword of him that sat upon the horse, which *sword* proceeded out of his mouth: and all the fowls were filled with their flesh. (Revelation 19:11-21)

First, Jesus Christ, the Righteous Judge, the King of Kings and Lord of Lords, the Messiah, will appear. He will come to earth riding on a white horse. Along with Him will be His Bride –the Believers whom He raptured from the earth, prior to the occurrence of the Great Tribulation on the earth. Also, coming with Him will be the armies of heaven that will also ride on white horses.

Second, on the earth, the Beast–Antichrist along with the False Prophet and the Gentile Confederate armies of the world will await His arrival to the Great Battle of God Almighty. These armies, under the supervision and leadership of Satan and his Antichrist/Beast and False Prophet, The King of kings and Lord of lords, the Messiah and Righteous Judge, will defeat them by the power of His Word.

Third, the King of kings will have His angels capture, chain, and cast Satan in the Bottomless Pit/Abyss and imprisoned him for 1000 years (Revelation 20:1-3)

Fourth, the Israeli people will repent of their sins, accept Jesus Christ as their Messiah, and receive all the promised-blessings. They will live in the land in peace. They will be at war anymore. The Lord will **"be the King over (them) all the earth"** (Zechariah 14:9).

Fifth, the King of Kings and Lord of Lords will set up the Millennial Kingdom on the earth. All His Saints will live with Him in the Kingdom. They will rule and reign with Him on the earth over the nations for 1000 years (2 Timothy 2:12; Revelation 5:10; 20:4 &6).

Event No. 4: The Final War- Gog & Magog II

> 7. And when the thousand years are expired, Satan shall be loosed out of his prison,
> 8. And shall go out to deceive the nations which are in the four quarters of the earth, Gog, and Magog, to gather them together to battle: the number of whom *is* as the sand of the sea.
> 9. And they went up on the breadth of the earth, and compassed the camp of the saints about, and the beloved city: and fire came down from God out of heaven, and devoured them. (Revelation 20:7-9)

The Final War, Gog & Magog II will take place on the earth. After God releases Satan from his imprisonment, he will gather all the nations of the world to fight against the Camp of the Saints and the Beloved City (Revelation 20:9). All these people will have been living in peace with the Prince of Peace ruling, will make war with the Lord, the Camp of the Saints, and the Beloved City. God will send down fire from above and will consume them all. In fact, the fire that will come down from God out of heaven will just not consume those who tried to fight against the King of Kings. The fire will also consume the heavens and the earth as we know them.

Event No. 5: The Old Heaven and Earth Will Be Replaced

The Judgment of God which ends Satan's rebellion and his leadership in the Final War, Gog & Magog II will bring about the destruction of heaven and earth as we know them. Note what the Bible states when God showed up on His Great White Throne, **"and him that sat on it, from whose face the earth and the heaven fled away"** (Revelation 20:11). The fact that Revelation 21:1 states, **"And I saw a new heaven and a new earth: for the first heaven and the first earth were passed away; and there was no more sea."** This means that even before the judgments of Satan and the Great White Throne, the old heaven and earth will be no more. In fact, the people who will appear before the Lord at the Great White Throne will have no place to stand. Remember Peter's prediction, **"But the heavens and the earth, which are now, by the same word, are kept in store, reserved unto fire against the day of judgment and perdition of ungodly men** (2 Peter 3:7).

Event No. 6: God's Final Judgment on Satan

> And the devil that deceived them was cast into the lake of fire and brimstone, where the beast and the false prophet *are*, and shall be tormented day and night for ever and ever. (Revelation 20:10)

The Judgment of Satan will occur. Satan who will lead the leaders of the nations and their armies in the Gog and Magog War II, against the Beloved City and the Camp of the Saints, God will judge them all by committing them to the Lake of fire. This will be Satan and his followers' end as predicted in Revelation 20:7–10.

Concluding Thoughts

First, from the three sets of prophecies presented, those of the Old Testament prophets, Jesus Christ, and the New Testament apostles, there is a common thread of unity and authenticity. All the prophecies of the Last Days or End Time, or as Jesus stated, "end of the world," will display many strange and unusual events with signs. People will see them occurring in the sky, on land, in the waters, and with the weather. These strange and unusual signs are beginning to occur today.

Second, in response to these strange and unusual events, science is urging people to change the way they treat Planet Earth. Also, scientists predict that without people making an urgent change, Planet Earth will come to an end. This means they see signs of "End Time" for the earth and its people. Interestingly, scientists are using scientific research for their findings. They are not using the Bible. Yet, their predictions are much like those of Bible End-Time predictions.

Third, science may not care about the predictions of the Bible. Therefore, their scientists may not fully see the truth of the outcomes they give to the phenomenon of Global Warming/Climate Change. Their outcomes could be God beginning to fulfill the predictions of His prophetic Word predicted long ago. Also, as God fulfills the prophecies, strange and unusual events will occur in the sky, on the land, in the waters, and with the weather, as predicted.

Fourth, it was previously seen in the chapters of this book that presented the prophecies of the Old and New Testament, God partially fulfilled some of them. However, there are many others that He must also fulfill in the end time. As God fulfills these prophecies, the events will occur with strange and unusual signs throughout the world. In many cases, people will become horrified over the loss of animals and people in the sky, on the land, and in the waters. Also, there will be strange signs with the weather.

Fifth, signs such as flooding, fires, pestilences, famines, wars, and rumors of wars, earthquakes, hailstorms, etc. are predicted to occur in the end times, per the Bible. Many such unusual signs are occurring today. However, scientists are crediting Global Climate Change for the events. It appears that they are not able to see the prophetic value of these signs. So, does this mean their view of the unusual events to be wrong? It could be that the events they see as those of Global Climate Change be *End Time Signs*? And God is now beginning to fulfill His Prophetic Word. Let us now investigate!

Sixth, the above eventful signs are those that will take place in the Sky, on the lands of the earth, in the Waters, and with the Weather. However, for greater details, readers may gain the following books written by the author. These are: (1) THE BIBLE EXPLAINS THE CLIMATE CHANGE CONTROVERSY, (2) THE COMING GREAT TRIBULATION ON THE EARTH, (3) ARMAGEDDON: The Greatest Battle Ever, (4) THE ONLY TRUE AND RIGHTEOUS JUDGE, (5) THE LAST LEADER IS COMING, (6) THE MYSTERY: Humans Becoming Immortals, (7) GLOBAL WARMING OR GOD'S WARNING: A Prophetic Explanation for the Strange and Unusual Events/Signs in the Sky, on the Land, in the Waters, and with the Weather. These books have all the eventful signs in great detail, and they can purchase them on Amazon at a reasonable price.

SECTION SIX

Decisions To Think About

Multitudes, multitudes in the valley of decision!
For the day of the LORD is near in the valley of decision.
The sun and moon will be darkened, and the stars no longer shine.
The LORD will roar from Zion and thunder from Jerusalem;
the earth and the heavens will tremble. **(Micah 3:14-16)**

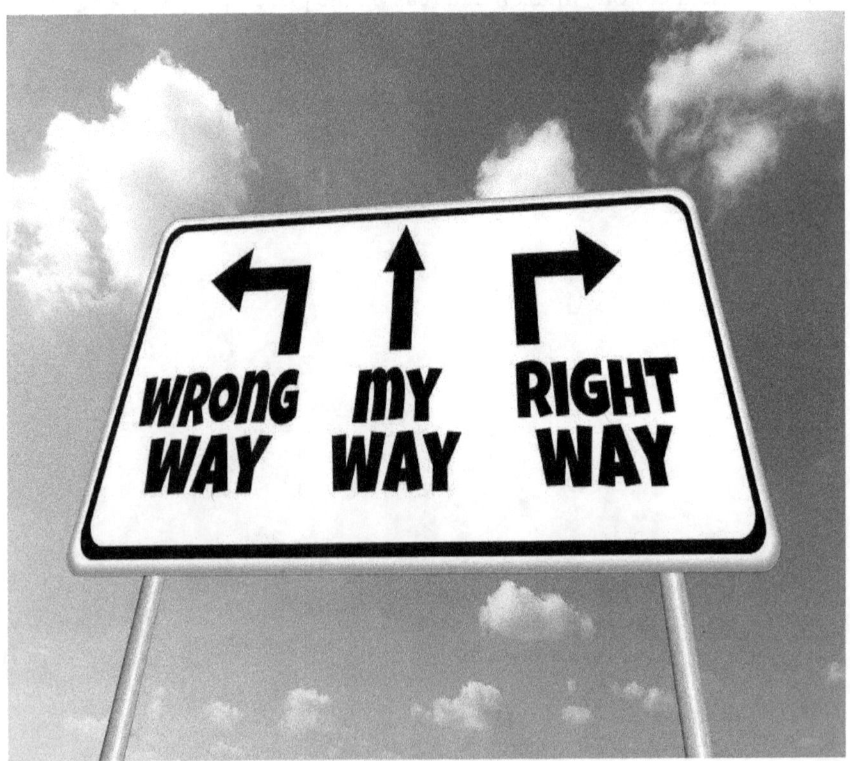

Which Way???

Chapter 21

CLIMATE CHANGE EVENTS ARE 'SIGNS OF THE END TIME?'

The Earth's Climate is Changing

Strange and unusual signs are taking place in the sky, on land, in the waters, and with the weather. So, to give answers to the chapter's big question, let us for a moment look back at the scientific findings and the Biblical prophecies to refresh our mental thinking. We need to really think about what we read about Global Warming-Climate Change and the Prophetic Messages of the Bible, God's Holy Word. The author believes that readers will understand the answer if Global Warming-Climate Change is a Warning from God. So, let us review what we have read.

The Readers' Review

The Review of the News Media Stories

You have read and you now know about the stories reported by the news media regarding strange and unusual activities taking place in the sky, on land, in the waters.

You now know that dead birds falling from the sky, science blamed the weather, probably because there were no other answers to give.

You now know about the cyber war eventful signs that are taking place in the sky, the stealing of e-mails and identification. On a national level, the United States recently placed sanctions on Russia for cyber interference in the 2016 presidential election. These activities are another form of war. Who knows what "buttons" someone will want to push as these cyberwars continue. The world is in a new frontier of fighting. No more bows and arrows, swords and shields, horses and horsemen, marching soldiers or tanks and rockets on the ground. It is with the flip of a switch or the pressing of a button that the entire world can come to a halt or annihilation.

You also know about the unusual activities on the land. You see how changes among people are taking place. The increase of hatred for others because of religion or the color of skin. Thus, there are all kinds of violence, kidnapping, pirating, terrorism, uprisings, conflicts, and wars. This planet is one of war.

You also know about the present display of selfishness among humans. There is no more sign of real love shown to each other. People are merciless, unconcerned, and are looking out only for themselves.

You now know about the problems in the world. Leadership is failing, the economic systems of nations are failing, and the world is looking for a world leader to solve their problems and heal the wounds of the world.

You also know about the diseases for which there are no cures. Besides the new onset of viruses and diseases, other new plagues are increasing, and there is no means of getting rid of them. Many of the old diseases like HIV/AIDS and many others do not have any concrete cures yet. Cancer is still a killer, and there is no real cure. Adding to these diseases is Coronavirus, the new killer.

You also know that the people of the world want answers for the unusual signs that have been taking place because they want to understand what is going on. They are looking to various sources for an answer, but there seems to be none coming that satisfies them. Many of them have given their interpretation through social media. Some believe that God is sending forth judgment, while others believe that these are acts of Mother Nature. The strange and unusual signs in the sky, on the land, in the waters, and with the weather, cause some people to be confused or frightened, and some are unconcerned.

The Review of the Scientific Findings

You have read and now know about global warming/climate change that its effect is a threat to the regions of the world. Also, it is a biological danger to humans and animals. This phenomenon will affect people's health, their food, and their water. It will affect world economics. All these areas are in danger of significant problems. Also, you have read about the effect of global warming/climate change on the military might of the world.

You have seen how the United States, Canada, and Great Britain have adjusted the military to fight the effect of the changing climate and the warming of the planet. In the year 2016, under the leadership of President Obama, the world leaders formed a pact to fight climate change. This president saw danger for the world.

You now know of the prediction for the sea and that ocean life is on the verge of extinction. Because of the heating of the oceans, the increase of CO2 gas is decreasing the levels of oxygen in the waters. You have also read about the great effect that global warming/climate change has on the weather. Science predicts the extreme weather conditions of earthquakes, tsunamis, hurricanes and tornadoes, severe flooding, and extreme firestorms, etc. caused by global warming/climate change. Many signs are taking place in the world and are on the rise and will continue on this track. However, there seem to be no real answers to the many questions people are seeking. Because many people are not satisfied with science's answers, therefore, they are still looking for answers. As a result, people still have various ideas on why there are strange and unusual signs occurring in the world.

The Review of the Biblical Prophetic Word

You also know the prophetic explanation for the strange and unusual signs taking place in the sky, on land, in the waters, and with the weather. The Old Testament prophets prophesied about such signs. Also, Jesus Christ prophesied about the coming earthquakes and weather, famine and pestilence, and the pain that people will experience. He mentioned the wars and rumors of wars, and these will only be the beginning of pain on people. People make wars, they fight in wars, and they die in wars. Today, wars are causing much pain to humankind—the killing, the running for survival, the sickness in trying to escape and living in camps, and the rejection of other nations in receiving refugees. If all these horrible activities are only the beginning of pain, that means that there are even greater eventful signs to befall the people of this earth. However, Jesus Christ said that these prophecies are signs of his return to the earth.

Also, you know about the predicted conditions that Jesus Christ gave to the people in the world before he returns to the earth. They will be like the people of Noah's day. They will be more interested in parties, marrying and giving in marriage. They will have no time for God. However, you read how many deceivers will come, and many of even the believers in Jesus Christ will become weak in their faith. Some will betray each other and turn them over to the authorities. However, there will be those who will carry on the gospel of eternal life.

Consider the story of the *fig tree*—how it blooms as an indicator that summer is near. You understood that the signs that are present in earthquakes, storms, famine, and pestilence are only the signs of the fig tree. Jesus Christ is soon to come.

You now know the prophecies of the New Testament apostles and how they also prophesied about activities in the sky, on land, in the waters, and with the weather. Mark and Luke prophesied about the Second Coming and the signs related to them. Luke, who was also a physician, prophesied about the conditions of people and how the strange and unusual signs will affect them physically, mentally, and psychologically. For example, the roaring of the sea will have a significant impact on the minds of people. The apostle John prophesied that Jesus Christ was coming back not only for his church at the rapture but also to return physically to the earth. You read that there will be The Great Tribulation on the earth, and all the strange and unusual signs associated with it.

You now know about the prophecies of the apostle Paul about the coming of Jesus Christ to take away his church. Also, how the condition of the people will be just before he returns. For example, he prophesied that people would turn away from God the Creator and serve and worship a beast, trees, and even a person. He mentioned that among the people there would be a change in their sexual orientation and practices. For example, he prophesied that men would have relations with men and women with women. Such relationships he prophesied are "against nature," meaning that the sexual apparatus that God gave to a woman and a man is different. Therefore, the prophecy stated that God has turned from such a reprobate mindset. In fact, the apostle prophesied that the time before the return of Jesus Christ would be "perilous times."

Also, you have read and now know how Jesus Christ will establish a kingdom on earth when he returns to the earth, where he will reign with the believers. Also, just before this reign starts, He will resurrect the tribulation saints and the Old Testament saints. The resurrection of two groups of Saints will complete the "First Resurrection Program.“ In the First Resurrection Program, (1) Jesus Christ's resurrection was first, (2) the second group will be the Believers of Jesus Christ at the Rapture - taking them from the graves and together with the living ones, transformed them and together take them to heaven, (3) the third group of Saints will be the resurrection of the Tribulation Saints, and (4), the resurrection of the Old Testament Saints (Daniel 12:1-3).

During the reign of Jesus Christ on earth, all the resurrected Saints will be in the Kingdom. At which time God will imprison Satan in the bottomless pit for one thousand years. After a thousand years of imprisonment, God will release him. He will go out and convince the Gentile nations to fight against Jesus Christ and the millennial kingdom. Gog and Magog and the prince of Russia will lead the nations in the war. God will defeat the armies, and he will cast Satan into the lake of fire with the beast and the false prophet. But this is not the end of all things.

There is a Second Resurrection coming. Also, the wicked angels who followed Satan, God will also judge.

You also know that after The Great White-Throne Judgment, hell and the grave will give up the dead. The persons that will come forth in the Second Resurrection are doomed to the lake of fire and brimstone, which the Bible describes as the Second Death. Also, there will be the judgment of the angels, the Gentile nations, and the nation of Israel. All these are the ending of the old-world form of government and the beginning of the new world Theocratic government Jesus Christ will establish. He will present to the Saints of God the new heaven, the new earth, and New Jerusalem- PARADISE. You know that when this Theocratic, Everlasting Paradise is presented, the Jews will be on the earth, and the believers of Jesus Christ will be in New Jerusalem with Jesus Christ. However, there will be communication between New Jerusalem and the New Heaven and Earth. In the last chapter of the book of Revelation, you read how God, in His mercy, today extends his love to humankind. He offers his final invitation right now to you. You can accept him now.

The Readers' Reflection

The above Review is a collection of facts the book presents to the readers. The news media, the scientific community, and the prophetic Word of God, the Bible shared information that readers may reflect. So, after reading the information about Global Warming and its Climate Change phenomenon and the predictions of the Bible, this chapter closes with these questions:

1. What do you think about the science of Global Warming and Climate Change?
2. Was the phenomenon of the science of Global Climate Change made clear?
3. Do you see the relationship of the prophetic Word of God, its fulfillment of past events/signs, as a guarantee that the other predicted eventful signs will come to pass?
4. Do you believe that the predicted prophetic signs are like those science predicts to be outcomes of Global

Warming/Climate Change?

5. Why are scientific predictions like the predictions of the Bible?

The author believes that when people consider the Prophetic predictions of God's Word, the Bible, then they should see science as an echo of what God's Word long ago has predicted. Let us investigate!

The Scientific Findings and Projections

What Science Predicts about Global Warming-Climate Change

1. Science predicts global warming/climate change will have a global effect on Africa, Asia, Australia, New Zealand, Europe, Latin America, North America, the polar regions, and the small islands.
2. Global warming/climate change will bring about drought, famine, wars, earthquakes, roaring seas, and challenges with the oceans.
3. Global Warming/Climate Change will upset *Earth's Biological Balance* between humans and animals.
4. *Health and Social Systems* per the World Health Organization (WHO) stated that global warming/climate change will cause large-scale environmental hazards to human physical, mental, psychological, and social health.
5. Science predicts the *Rise of Diseases in the World.* Global warming/climate change has exposed human beings to weather patterns that affect temperature, precipitation, sea-level rise, and more frequent extreme eventful signs.
6. Global warming/climate change will *Increase the Conflict and Migration of People.* It will displace millions of people from the shoreline areas to other areas because of shoreline erosion or severe drought.
7. Global warming/climate change will cause a *Decrease in World Economics and Influence on the Military Might of the World.*

8. There will be wars when the economic systems fail and there are shortages of food and water.
9. Increase in storms such as earthquakes, tsunamis, flooding, increased heating, etc.

The Biblical Prophetic Predictions

What The Bible Predicts about End of the World Signs

First, Jesus Christ's teachings and His prophecies about strange and unusual signs that will occur in the sky, on land, in the waters, and with the weather. Jesus Christ predicted that in these four mentioned areas, the events will be signs of the ***last days/end-time*** just before He returns to the earth to set up His Kingdom. Note the events or signs He predicts:

1. Storms, such as earthquakes
2. There will be famines and pestilence
3. Wars and rumors of wars
4. People will hate each other, and even the people who claimed to be believers and should follow Him will turn against each other and deliver some to death
5. The systems of government will turn against believers, but the Word of God and the gospel message will go to all corners of the earth
6. The nation of Israel will have much tribulation, and
7. There will be a tribulation across the land just before He returns to the earth.

The Concluding Statements about Global Warming – Climate Change and the Prophetic Predictions of the Bible

1. It would appear that what science calls the effects of global warming/climate change are the same eventful signs that Jesus Christ predicts to take place in the sky, on the land, in the waters, and with the weather.
2. Because science's predictions on global warming/climate change are like the predictions of the Bible, it may be that we

are already living in the *'End Times'* right now.

3. Both the prophecies of the Bible and the findings of science appear to be pointing in the same direction. It appears that the scientific predictions and the Holy Bible predictions share many similarities.
4. The Holy Bible is the Word of God. Scientists are who are the creation of God. God made them in His image.
5. Real knowledge comes from God, and true wisdom, the Holy Bible tells us.
6. The knowledge and understanding of scientists are gifts from God. So, their discoveries about global climate change and what it will do to the earth, people, and animals are only supporting the Word of God.
7. Therefore, Global Climate Change could be *an instrument* that God is using to warn people of the *End of the World*, as Jesus Christ states in Matthew 24.
8. It seems clear and safe to say—since *the scientists are not sharing information with God or advising God—that it must be God, the Holy One, in His Divine wisdom, who is advising the scientists.*
9. Therefore, it is also safe to say that (1) long before there were scientists with their knowledge, there was God. (2) Also, before scientists, there were the prophetic predictions from the Old Testament prophets, Jesus Christ, and many of the New Testament apostles. (3) Because Jesus Christ's coming back to the earth is near, God is allowing scientific knowledge to humankind, as predicted by the prophet Daniel, "even to the time of the end: many shall run to and fro, and knowledge shall be increased" (Daniel 12:4).

So, God will continue to enlighten scientists about His universe, global warming/climate change, and the activities in the sky, on land, in the waters, and with the weather. However, because God Almighty predicted such findings long ago, it seems conclusive that the climate change activities and predictions by scientists are predictions concerning "End Time Signs" that God is allowing to take place. See, God is using these signs to warn people of the "End Time." Also, He wants humankind to prepare themselves to escape the coming damnation.

The author concludes that after testing the information about global warming/climate change, there is a clear conclusion:

- Global Warming Climate Change is real.
- Global Warming Climate Change is taking place now and will continue until Jesus Christ returns to the earth.
- Global Warming Climate Change Events are **'signs of the last days/End Time'** that God is using to warn people of the Coming King of Kings and Messiah. So, be ready to meet Him soon.
- So, Why is God Warning people about the phenomenon of Global Warming – Climate Change?

Chapter 22

WHY IS GOD USING STRANGE AND UNUSUAL SIGNS?

The apostle Peter made it clear in 2 Peter 3, that in the face of disbelief by some people. *First,* we *must understand* that in the *End Times,* some people will continue to degrade and make fun of the prophecies of the Word of God. Some people will also debate and contradict the Word of God concerning the coming of Jesus Christ back to the earth, and the signs of the *End Times*. However, in their ignorance, they will deliberately forget that long ago by God's word, He made the heavens and formed the earth. Also, that *right now*, by the same word of His power, He will destroy the heavens and earth by fire. On such a day he will not only bring about this judgment but will destroy the ungodly.

Second, although this promise of God has been in His Word for many years, people must remember, with the Lord a day is like a thousand years, and a thousand years are like a day. Also, the Lord is not slow in keeping his promise, as some understand slowness. Instead, He is patient with people, not wanting anyone to perish, but everyone to come to repentance.

Third, the reason God is being patient with people of the earth, one day the Lord will come like a thief. On that day, the heavens will disappear with a roar; a fire will destroy the elements, and the earth and everything in it will disappear. Since God will destroy everything in this way, this should be a warning to people to live holy and godly lives as each person looks forward to the day of God. However, the Believers of Jesus Christ, in keeping with the promise of this day of the destruction of the heavens by fire, and the melting of the elements or atoms of the universe, in keeping with this promise should look forward to a new heaven and a new earth, where righteousness dwells.

God Wants All People to Know that His Prophetic Word Is Truth

The Holy Bible is the book of all the books, the study of all researchers. It is the work of God and not of men. It defends itself, for it is not dead. Every word is alive and can make humankind alive. The prophecies are valid, authentic, trustworthy, and dependable. For example, in defending its prophecies, it states:

We have also a surer word of prophecy; whereunto ye do well that ye take heed, as unto a light that shineth in a dark place, until the day dawn, and the day star arise in your hearts: Knowing this first, that no prophecy of the scripture is of any private interpretation. For the prophecy came not in old time by the will of man: but holy men of God spake as they were moved by the Holy Ghost. (2 Peter 1:19– 21)

The Prophecies of the Bible are Anchors for the Word of God

First, the above verses speak of the prophecies of the Holy Bible. This scriptural reference authenticates the Word. It reminds humankind that the prophetic messages are dependable, and people must give attention to them. It is through these prophetic messages that the power of light shines within the hearts of humankind, in the darkness of night where others will never see. Moreover—and above all—people must understand that prophecy has no beginnings or origins with humans. It is the Holy Spirit of God that reveals to the person the things of God that must come to pass. Therefore, the people of God wrote as the Holy Spirit of God led them.

Second, reading the prophecies from the Holy Bible and seeing that many of them God has already fulfilled, people cannot help but understand that these prophecies are the actual Word of a Holy God. The prophecies of God's Word are accurate and trustworthy, and He will fulfill them. Look at the predictions of Jesus Christ concerning his past, present, and future activities that are in the prophecies of the Holy Bible. If you look at the prophecies carefully, you will see that when a prophet foretold a future event, God fulfills the promise. Therefore, those prophecies that the Old Testament prophets, the New Testament apostles, and Jesus Christ predicted in God's Word, to occur in the "End Time' He will also fulfill.

Third, the last Book of the Bible, the Book of Revelation, the apostle John reminds all people with these predictions:

> "These words are trustworthy and true. The Lord, the God who inspires the prophets, sent his angel to show his servants things that must soon take place."
> "Look, I am coming soon! Blessed is the one who keeps the words of the prophecy written in this scroll."
> "Do not seal up the words of the prophecy of this scroll, because the time is near." (Revelation 22:6,7, &10).

Because of God's love to and for people, He is warning with the phenomenon of global climate change and its signs/events in the sky, on the land, in the waters, and with the weather. At the beginning of this chapter, the apostle Peter warns people that the Lord is not slow in keeping his promise, as some understand slowness. Instead, He is patient with all people, not wanting anyone to perish, but everyone to come to repentance (see 2 Peter 3:9).

Chapter 23

God Gives the Final Decision to all People – What is Your Decision?

Life?

"And the Spirit and the bride say, Come. And let him that heareth say, Come. And let him that is thirsty come. And whosoever will let him take the water of life freely" (Revelation 22:17).

Death?

"Wherefore, as by one man sin entered into the world, and death by sin; and so, death passed upon all men, for that all have sinned" (Romans 5:12)

The apostle *Paul* sums up humankind's condition with this scripture from Romans 3:

> There are no righteous people, no, not one: There is none that understandeth, there is none that seeketh after God. They are all gone out of the way, they are together become unprofitable; there is none that doeth good, no, not one. Their throat is an open sepulcher; with their tongues, they have used deceit; the poison of asps is under their lips: Whose mouths are full of cursing and bitterness: Their feet are swift to shed blood: Destruction and misery are in their ways: And the way of peace have they not known: There is no fear of God before their eyes. (Romans 3:9–18)

Listen to how God feels about humans and the message he gave to the prophet *Ezekiel.* His message from God still stands today and tomorrow, until he comes:

> Say unto them, As I live, saith the Lord GOD, I have no pleasure in the death of the wicked; but that the wicked turn from his way and live: turn ye, turn ye from your evil ways; for why will ye die, O house of Israel? (Ezekiel 33:11).

Jesus Christ has completed the work of salvation. People now must make up their minds to come back to God. He does not care what a person has done or where he or she has been. He wants everyone to come to him. Listen to what God is saying through the prophet Isaiah. "Come now, and let us reason together, saith the LORD: though your sins are as scarlet, they shall be as white as snow; though they are red like crimson, they shall be as wool" (Isaiah 1:18).

Note, John 3:16 states:

> "For God so loved the world that he gave his only begotten Son, that whosoever believeth in him shall not perish but have everlasting life (John 3:16).

God instructed people through His Word, the Bible, how they may get Everlasting Life. The apostle Paul explains it in a Book in the Bible – Romans. He outlines four states in Chapter 10:8–13. Note the stages:

(1) That if thou shalt confess with thy mouth the Lord Jesus, and shalt believe in thine heart that God hath raised him from the dead, thou shalt be saved.
(2) For with the heart man believeth unto righteousness, and with the mouth confession of sins yields salvation.
(3) For the scripture saith, Whosoever believeth on him shall not be ashamed. For there is no difference between the Jew and the Greek: for the same Lord over all is rich unto all that call upon him.
(4) For whosoever shall call upon the name of the Lord shall be saved. (Romans 10:8–13)

Any person who calls upon the Name of Jesus Christ will be saved and have eternal life. So, consider Jesus now and make Him your choice. He wants to give you and all the people the water of life. For another time listen:

> **And the Spirit and the bride say, Come. And let him that heareth say, Come. And let him that is thirsty come. And whosoever will let him take the water of life freely** (Revelation 22:17).

Jesus Christ is saying, **"Surely I come quickly."** Also, when people know Jesus Christ as Lord and Savior, they can say, **"Amen. Even so, come, Lord Jesus"** (Revelation 22:20).

Settle Your Account Now! God Bless You!

Brother Des (Isaiah 41:10).

Tables

Table No. 1

See the Wikipedia Unemployment rate by country online
https://en.wikipedia.org/wiki/List_of_countries_by_unemployment_rate

Table No. 2

Earthquakes and Tsunamis (see the link below for online statistics). (http://en.wikipedia.org/wiki/Lists_of_earthquakes#Deadliest_earthquakes_on_r record; & CNN)

Table No. 3

Wars (see the link below for online statistics). https://ca-69boston.org/death-tolls-revisited/

****

Appendices

Appendices Nos. 1-9 – Animal Mass Deaths Worldwide (see the link below for online statistics).
http://www.end-times-prophecy.org/animal-deaths-birds-fish-end-times.html

References

Abadie, A. (2004). "Poverty, Political Freedom, and the Roots of Terrorism." Retrieved from http://www.nber.org/papers/w10859.

ABC News, Tuesday, February 19, 2013. "Russian Meteor: Rushing to Cash-in on the Blast." Retrieved from http://abcnews.go.com/International/russian-meteor-rushing-cash-blast/story?id=18522807.

Ackerman, S. (2013). WIRED. "There is no way to stop space rocks from hurtling to earth and killing you." Retrieved from https://www.wired.com/2013/02/russia- meteor/.

AIR WORLDWIDE. (2010). European Windstorms: Implications of Storm Clustering on Definitions of Occurrence Losses." Retrieved from http://www.air-worldwide.com/Publications/AIR-Currents/2010/European-Windstorms-Implications-of-Storm-Clustering-onDefinitions-of-Occurrence-Losses.

AIR WORLDWIDE. "The Anatomy of a Tornado." Retrieved from http:// www.air-worldwide.com/Facet-Search/Search-Results/.

Alesina, A., S. Ozler, R. Roubini, and P. Swagel. (1992). "Political Instability and Economic Growth Dash." Harvard. Retrieved from https://dash.harvard.edu/bitstream/handle/1/4553024/alesina_instabilitygrowth. pdf?sequence=2.

Aljazeera. (2019). "Iran's protests: All you need to know in 600 words." Retrieved from, https://www.aljazeera.com/news/2019/11/iran-protests-600-words- 191118060831036.html

Alster, P. (2015). "Potentially game-changing oil reserves discovered in Israel." Retrieved from http://www.foxnews.com/world/2015/10/07/potentially-game- changing-oil-reserves-discovered-in-israel.html.

amfAR Making AIDS History (2015). Retrieved from http://www.amfar.org/worldwide-aids-stats/).

Associated Press. (2013 Report). "Blizzard Drops More Than 2 Feet of

Snow on Northeast." Retrieved from
http://abcnews.go.com/US/ blizzard-2013-fierce- storm-drops-feet-snow-northeast/story?id=18443349.

Associated Press, Utah. (2011). "Thousands of Birds Make a Crash Landing in Utah." Retrieved from
http://news.yahoo.com/thousands-birds-crashlanding-utah-152219107.html.

Associated Press. (2011). "Thousands of migratory birds make crash landing in Utah, killing about 1,500 of them." Retrieved from
http://www.
syracuse.com/news/index.ssf/2011/12/thousands_of_migratory_birds_m.html.

BBC News. (2014). "Large numbers of baby gannets or 'gugas' found dead in East Lothian." Retrieved from
http://www.bbc.co.uk/news/uk-scotland- edinburgh-east-fife-29309309.

BBC News. (2011). "United Kingdom. Iceland volcano: Grimsvotn eruption hits flights." Retrieved from
http://www.bbc.co.uk/news/worldeurope-13489944 (BBC News).

Becker, G. and R. Posner. (2008). "Terrorism and Economic Development – Posner's Comment." Retrieved from
http://www.becker-posner-blog. com/2008/01/terrorism-and-economic-development--posners-comment.

befoundalive. (2016). "WATER: The Staff of Life." Retrieved from
http://befoundalive.blogspot.com/2010/10/water-staff-of-life.html

Bindoff, N. L., J. Willebrand. (2007). "Observations: Oceanic Climate Change and Sea Level." Retrieved from
http://eesc.columbia.edu/courses/
v1003/readings/IPCC%20AR4%202007/AR4WG1_Pub_Ch05.pdf.

BING.COM. (2015). "Floods in Canada." Retrieved from
https:// www.bing.com/search?q=flooding%20in%20canada%20

Blake, E. S., E. N. Rappaport, & C. W. Landsea. (2007). "The Deadliest, Costliest, and Most Intense United States Tropical Cyclones from 1851 to 2006 (and Other Frequently Requested Hurricane Facts)." Retrieved on 2008-03-19 (www.nch.noaa.gov/NWS-TPC-5.odf).

Boutilier, A. (2014). The Star, Canada News. "Canadian Forces watching climate change carefully, Rob Nicholson says." Retrieved from https://www.thestar.com/news/canada/2014/08/26/canadian_forces_watching_ climate_change_carefully_rob_nicholson_says.html.

Bowater, D. (2014). "Record floods in Brazil bring chaos to Amazon towns." Retrieved from BBC News. http://www.bbc.com/news/ world-latin-america- 28123680.

Brannon, M. (2015). "Obama's Decision to Take on Refugees Faces Criticism." CBS. Retrieved from http://www.cbsnews.com/news/ obama-decision-refugees- fleeing-syria-civil-war-criticism/.

Brocchetto, M. (2012). "2,300 birds found dead along Chilean beaches." CNN. Retrieved from http://www.cnn.com/2012/05/12/world/americas/ chile-peru- dead-birds/.

Brzoska, M. (2012). "Climate change and the military in China, Russia, the United Kingdom, and the United States." Retrieved from, https://www.tandfonline.com/doi/abs/10.1177/0096340212438384

Bryner, J. (2008) LiveScience. "Bad Habits: Why We Can't Stop." Retrieved from http://www.livescience.com/1191-bad-habits-stop.html.

Caballero, R. J. & A. Simsek. (2009). "Complexity and Financial Panics." Retrieved from http://www.nber.org/papers/w14997.pdf.

Caldwell, Z. (2010). Retrieved from http://www.ipsnews.net/news. asp?idnews=48739).

Callahan, M. (2014). "Large-scale die-off small seabird along Sonoma Coast." The Press Democrat. Retrieved from http://www.pressdemocrat.com/news/3145997-181/large-scale-die-off-of-small-seabird.

Carroll, R. (2010). National Geographic News. "Stormier Arctic Predicted as Ice Melts." Retrieved from http://news.nationalgeographic.com/ news/2009/10/091015-arctic-ice-free-gone-storms.html Sahney, Benton, & Falcon-Lang, 2010).

CBC News, 2016). Retrieved form http://www.cbc.ca/news/canada/newfoundland-labrador/states-of-emergencies-rainstorm-aftermath-1.3799472.

CBC News, British Columbia. (2015). "Cassin auklets found washed up near Tofino." Retrieved from http://www.cbc.ca/news/canada/british-columbia/cassin-auklets-found-washed-up-near-tofino-1.2891409.

Censky, A. (2020). n p r news. "Heavily Armed Protesters Gather Again at Michigan's Capitol." Retrieved from https://www.npr.org/2020/05/14/855918852/heavily-armed-protesters-gather-again-at-michigans-capitol-denouncing-home-order

Chika, U.P. (2011). "Strategies for Climate Change Adaptation Among Rural Households in Imo State, Nigeria." Department of Agricultural Extension, University of Nigeria, Nsukka. Retrieved from http://repository.unn.edu.

Church of the Eternal God. "Booklet, Middle Eastern and African Nations in Bible Prophecy." Retrieved from http://www.eternalgod.org/booklet-2187/.

Clarke, R. & Knake. "Cyber War—The Next Threat to National Security and What to About It." HarperCollins, 2010.

Clapper, J. R. (2015). "Worldwide Threat Assessment of the US Intelligence Community." Senate Armed Services Committee, the United States Retrieved from https://en.wikipedia.org/wiki/Cyberwarfare.

Climatology Geographic Site. (2015). "The floods are more frequent than they used to be?" Retrieved from http://en.climatologiageografica.com.br/station-flood-frequently-that-used-be/.

Congressional Budget Office Report. (2011). Retrieved from https://www. cbo.gov/publication/21999.

Conners, D. (2012). "Are large earthquakes increasing in frequency?" Retrieved from
EarthSky, http://earthsky.org/earth/ are-large-earthquakes-increasing-in-frequency.

Council on Foreign Relations (2002). Interview: "Greenburg: Global Implications if IAG Fails." Retrieved from http://www.cfr.org/world/greenberg-global-implications-if-aig-fails/p17252.

CNN. (2015). "South Carolina governor calls deadly rain a 'thousand-year' event." http://www.cnn.com/2015/10/04/us/east-coast-rain-flood/index.html.

Crowley, T. J.; G. R. North. (May 1988). "Abrupt Climate Change and Extinction Events in Earth's History." Science 240 (4855): 996–1002.

CTV Winnipeg (2014), "Recent storms cover beach in dead birds." Retrieved from http://winnipeg.ctvnews.ca/ recent-storms-cover-beach-in-dead-birdTs- 1.1898801.

Davies, R. (2015). floodList. "Paraguay – 8,000 Families Evacuated in Asuncion after Paraguay River Overflows." Retrieved from http://www.nasa.gov/mission_pages/fires/main/modis-10-overview.html.

Davies, R. (2015). floodList. "Guinea – 450 mm of Rain in 3 Days - Floods in - Conakry Leave 4 Dead." Retrieved from http://floodlist.com/africa/ guinea- 450mm-rain-floods-conakry.

Davies, R. (2015). floodList. "15 Dead after Floods and Landslides in Northern Vietnam." Retrieved from http://floodlist.com/ asia/15-dead-after-floods-and- landslides-in-northern-vietnam.

Davies, R. (2015). "NCAR Scientists to Tackle Mystery of Nighty Thunderstorms." Retrieved from NCAR Scientists, http://floodlist.com/protection/ncar-scientists-to-tackle-mystery-of-night-thunderstorms.

Degregory, P. & D. K. Li. (2015). "Day care workers charged with running toddler 'Fight Club'". The New York Post. Retrieved from http://nypost.com/2015/09/01/ day-care-workers-had-young-kids-brawl-in-fight- club-videos/.

DEL PERU PARA ELMUNDO (2014). "Some 1,832 guaneras birds were found dead in the coast of Arequipa." Retrieved from (https://translate.google.com/translate?sl=auto&tl=en&js=y&prev=_t&hl=en&ie=UTF-8&u=http%3A%2F%2Fwww.

DeMarban, A. (2016). Retrieved from http://www.adn.com/article/20131126/hundreds-dead-seabirds-wash-ashore-alaska-islandbering-sea).

De Moura, H. (2011) CNN "Death toll Brazil flooding continues rising." Retrieved from http://edition.cnn.com/2011/WORLD/americas/01/17/brazil.flooding/index.html.

Desanker, et al, (2001) "It has been argued that environmental degradation, loss of access to resources (e.g., water resources)."

Desanker, et al, (2001). "Economic impacts of climate change." Wikipedia.

Dastagir, A. E. (2017) "The State of hate in America: In an America deeply divided, hate incidents appear to be increasing and growing more brutal." USA TODAY. 7.10.17.

Doherty, T. J. & S. Clayton. (2011). "The Psychological Impacts of Global Climate Change." Retrieved from http://www.apa.org/pubs/journals/ releases/amp-66-4-265.pdf.

Dunne, John P., Ronald J. Stouffer, and Jasmin G. John (2013). "Heat stress reduces labor capacity under climate warming". Geophysical Fluid Dynamics Laboratory. Bibcode:2013NatCC…3..563D.doi:10.1038/nclimate1827.

DW (2013). Retrieved from http://www.dw.com/en/ floods-in-germany-a-sign- of-climate-change/a-16860917.

Earth Changing Extremities (2016). "Thousands of whitefish dead in eastern Idaho, USA." Retrieved from https://yamkin.wordpress.com/2016/09/16/ thousands-of-whitefish-dead-in-eastern-idaho-usa/.

Easterling; et al. (2007). Climate Change 2007: Working Group II Impacts, Adaptation and Vulnerability. "Chapter 5: Food, Fibre, and Forest Products", 5.4.1 Primary effects and interactions Missing or empty |title= (help), in IPCC AR4 WG2 2007, p. 282.

Emanuel, K. (2013). "Downscaling CMIP5 climate models show

increased tropical cyclone activity over the 21st century." Retrieved from CrossMark, http://www.pnas.org/content/110/30/12219.abstract.

End Time Prophecy, (2013-2019) "Mass Animal Deaths for 2017." Retrieved from, http://www.end-times-prophecy.org/animal-deaths-birds-fish-end- times.html

Farbotko, C. (2012). "Climate change, migration and adaptation in Funafuti, Tuvalu." Journal of Global Environment Change. Retrieved- http://www.academia.edu/15472334/ Climate_change_migration_and_adaptation_in_Funafuti_Tuvalu.

Fischlin A., et al. "Ecosystems, their properties, goods, and services." In Climate change 2007 climate change impacts, adaptation, and vulnerability. Fourth Assessment Report of the Intergovernmental Panel on Climate Change, pp. 211–272. Cambridge, UK: Cambridge University Press, 2007.

floodList - CARBON BRIEF IN USA. (2015). "How Storm Surges and Heavy Rainfall Drive Coastal Flood Risk in the US." Retrieved from http://floodlist.com/america/usa/storm-surges-heavy-rainfall-drive-coastal-flood-risk.

floodList - CARBON BRIEF IN USA. (2015). Retrieved from http:// floodlist.com/america/usa/3-killed-flash-floods-ripley-ohio-july-2015.

floodList - CARBON BRIEF IN USA. (2015). "Model Helps City Planners Prepare for Weather Floods and Storms." Retrieved from http://floodlist.com/protection/model-city-planners-prepare-floods-storms.

Focus Taiwan (2015). "COA introduces new poultry farming regulations as bird flu spreads." Retrieved from http://focustaiwan.tw/news/ aeco/201503040026.aspx.

Fogarty, D. (2011). "Scientists see climate change link to Australian floods." Retrieved from REUTERS SCIENCE NEWS, http://www.reuters.com/article/us-climate-australia-floods-idUSTRE70B1XF20110112#CDmzO JYywziXwxPV.97.

Fox, C. (2015). "Hawaii Just Got Hit by A July Snowstorm (Seriously)." Retrieved from http://www.huffingtonpost.com/entry/hawaii-summer-snow- storm_us_55a973dde4b065dfe89e737f.

Fox2Now, St. Louis. (2012). "Flock of birds falls from the sky in Springfield, MO." Retrieved from http://fox2now.com/2012/11/20/ flock-of-birds-falls-from- the-sky-in-springfield-mo/.

Francis, J. A. & Vavrus, S. J. (2012). "Evidence linking Arctic amplification to extreme weather in mid-latitudes." Retrieved from, https://agupubs.onlinelibrary.wiley.com/doi/full/10.1029/2012GL051000

Freeman, D. (2010). "The missing link: China, climate change and national security." Asia Papers 5 (8). Brussels Institute for Contemporary China Studies. Retrieved from, http://www.vub.ac.be/biccs/site/asserts/files/papers/Asia%20papers/2010 12%20%20Climate%20Change%20Security.pdf.

France-Presse, A. (2016). "Vietnam investigates mass fish deaths." Retrieved from https://www.theguardian.com/environment/2016/apr/21/vietnam-investigates-mass-fish-deaths-pollution.

Gaines, J. (2011) Climate Management Issues: Economics, Sociology, and Politics, CRC Press. Climate Management Issues: Economics, Sociology, and Politics.

GARDENS FOR HEALTH INTERNATIONAL. (2015). "Gardens for Climate Part I: Let's Cross Sectors." Retrieved from http://www.gardensforhealth. org/building-resilience-to-climate-change-lets-cross-sectors-2/.

Gearin, M. (2013). "Climate change makes super typhoons worse, says UN meteorological agency." Retrieved from ABC NEWS. http://www.abc.net.au/news/2013-11-14/climate-change-making-super-typhoons-worse/5090724.

German, G. (2013). "Atmospheric Oxygen Levels Are Dropping Faster Than Atmospheric Carbon Levels Are Rising." Retrieved from (http:// disinfo.com/2013/01/atmospheric-oxygen-levels-are-dropping-faster- thanatmospheric-carbon-levels-are-rising/.

Gillette, B. (2017). "The Messiah's Gate." Retrieved from Rapture Notes. http://www.rapturenotes.com/eastern-gate.html

Gines, J. K. (2011). "Climate Management Issues: Economics, Sociology, and Politics." CRC Press, Taylor & Francis Group. Retrieved from https://www.crcpress.com/Climate-Management-Issues-Economics-Sociology-and-Politics/ Gines/p/book/9781439861066

Global Health Observatory (GHO) Data (2015). "HIV/AIDS." Retrieved from http://www.who.int/gho/hiv/en/.

Global News (2014). "Watch: Natural disasters make 2014 a year to forget for many." Retrieved from http://globalnews.ca/news/1746702/ watch-natural- disasters-make-2014-a-year-to-forget-for-many/.

Godfrey, A. (2012). "Thousands of birds drop out of the skies in England." The *Herald Sun*. Retrieved from http://www.heraldsun. com.au/travel/travel- news/thousands-of-birds-to-drop-out-of-the-skyin-england/story-

Goldenberg, S. (2010). "Climate Vulnerability Monitor 2010: The State of the Climate Crisis." Fundación DARA Internacional. Retrieved from https:// books.google.com/books?id

Goldenberg, S. (2010) The Guardian. "Environment." Retrieved from https://www.revealnews.org/author/suzanne-goldenberg/.

Gore, A. OUR CHOICE: A Plan to Solve the Climate Crisis. Rodale, Emmaus, PA., 2009.

Graham, R. (2011). "Up to 350,000 seabirds die in most deadly 'Wreck' in New Zealand's history" Retrieved from https://birdsnews.com/2011/first-surviving-broad-billed-prions-being-released-after-massive-july%E2%80%98wreck%E2%80%99-in-new-zealand/

Graves, L. (2013). Huffington Post. "Obama Climate Change 2013 Policy Speech Outlines Executive Orders." Retrieved from http://www.huffingtonpost.com/2013/06/25/obama-climate-change2013_n_3497151.html.

Henry, P. B. (2013) Global Leadership. Retrieved from http://www.

duq.edu/academics/schools/leadership-and-professional-advancement/ graduate-degrees/global-leadership?gclid=CKrNjI24n7wCFacDOgodpS AArQ).

Hoppe, P. (2013). reliefweb. "Floods dominate natural catastrophe statistics in first half of 2013." Retrieved from http://reliefweb.int/ report/world/floods-dominate-natural-catastrophe-statistics-first-half2013#sthash.3Gre7kNl.dpuf.

ICELAND REVIEW ONLINE (2012). "Cause of Death of Herring in West Iceland a Mystery Retrieved from http://icelandreview.com/ news/2012/12/17/cause-death-herring-west-iceland-mystery) Updated: January 30, 2014, 20:24.

ICC International Maritime Bureau. (2019). "Piracy and Armed Robbery Against Ships." Retrieved from https://iccwbo.org/media-wall/news-speeches/seas-off- west-africa-worlds-worst-pirate-attacks-imb-reports/

Intergovernmental Panel on Climate Change. (2007). Climate Change 2007: Working Group II: Impacts, Adaptation, and Vulnerability." Retrieved from http://www.ipcc.ch/publications_and_data/ar4/wg2/en/ ch19.html.

Intergovernmental Panel on Climate Change (IPPC) -Wikipedia). "Effects of global warming on human health." Retrieved from http://wikivisually. com/wiki/Effects_of_global_warming_on_human_health.

Intergovernmental Panel on Climate Change. (2007. "Observed Changes in climate and their effects." Retrieved from http://www.ipcc.ch/ publications_and_data/ar4/syr/en/spms1.html.

Jakobsen, L. (2010). "China prepares for an ice-free Arctic. SIPRI Insights on Peace and Security (2), March. Available at: http://books.sipri.org/files/insight/SIPRIInsight1002.pdf.

Kim, L. (2021 "Russian Protesters Demand Alexei Navalny's Release from Prison." Retrieved from Russian Protesters Demand Alexei Navalny's Release From Prison : NPR

LaMotte, S. (2016). "Zika has been sexually transmitted in Texas, CDC confirms." CNN Retrieved from http://www.cnn.com/2016/02/02/health/ zika- virus-sexual-contact-texas/index.html.

LIBERTARIANISM.org. (1999). "The Greatest Century That Ever Was: 25 Miraculous/Trends of the Past 100 Years. "Retrieved from https://www.libertarianism.org/publications/essays/greatest-century-ever-was- 25-miraculous-trends-past-100-years.

LiveScience (2006). "Peace or War? How Early Humans Behaved." Retrieved from http://www.livescience.com/640-peace-war-early-humans-behaved.html.

Longbottom, W. (2011). "Now 8,000 Doves fall dead from Italy's Skies. What in the hell is really going on here?"

Longshot's Blog (2011). "Now 8,000 Doves fall dead from Italy's Skies. What in the hell is really going on here?" Retrieved from https:// longshotsblues.wordpress.com/2011/01/08/now-8000-doves-fall-deadfrom- italys-skies-what-in-the-hell-is-really-going-on-here/.

Lovett, R. A. (2009). "North Magnetic Pole Moving Due to Core Flux." Retrieved from http://news.nationalgeographic.com/news/2009/12/091224-north-pole-magnetic-russia-earth-core.html#.

Manzo, D. & D. Chiu. (2015). "Catastrophic Flooding Throughout South Carolina." Retrieved from http://abcnews.go.com/US/charleston-south-carolina- soaked-worst-rains-1000-years/ story?id=34233408).

Marlborough Express (2012). "Salmon deaths a mystery." Retrieved from http://www.stuff.co.nz/marlborough-express/news/6564173/Salmon-deaths-a- mystery.

Martiner. (2014). "Chronology of Ebola Virus Disease Outbreaks – 19762016." Retrieved from http://healthintelligence.drupalgardens.com/ content/chronology- ebola-virus-disease-outbreaks-1976-2014.

Medecins Sans Frontieres (2016). Doctors Without Borders. Medical aid where it is needed most. Independent. Neutral. Impartial. Retrieved from http://www.doctorswithoutborders.org/

Met Office Exeter UK (2012). "Arctic sea ice 2012." Retrieved from http:// www.metoffice.gov.uk/research/news/2012/sea-ice-2012

Mercola (2011). "Mass Death of Birds and Fish: Is There a Cover-Up?" Retrieved from http://articles.mercola.com/sites/articles/archive/2011/01/25/the-10-leading-theories-for-dead-birds-and-fish.aspx.

Merriam-Webster's Collegiate Dictionary. Copyright 2000 by Merriam-Webster, Incorporated. The United States of America.

Meyer, R. (2015). "How Did Hurricane Patricia Intensify So Quickly?" Retrieved from The Atlantic, http://www.theatlantic.com/science/archive/2015/10/how- did-hurricane-patricia-intensify-so-fast/412246.

Michael, P. (2013). "Bird experts and scientists left puzzled as birds fall dead from the north Queensland skies." The *Courier Mail*. Retrieved from http://www.couriermail.com.au/news/queensland/bird-experts-andscientists-left-puzzled-as-birds-fall-dead-from-north-queensland-skies/story-fnihsrf2-

MIT Technology Review (2017). Ten Breakthrough Technologies for 2017. Retrieved from https://www.technologyreview.com/lists/ technologies/2017/.

Molina, B. (2019). "Scientists explain why we're just now learning about a giant meteor that exploded over Earth last year." Retrieved from, https://www.usatoday.com/story/news/world/2019/03/19/meteor-exploded-over-bering-sea-biggest-blast-since-2013-says-nasa/3211856002/

Morgan, E. (2011). "Hundreds of birds found dead in Davis County." Deseret News Utah Retrieved from http://www.deseretnews.com/article/705396148/Hundreds-of-birds-found-dead-in-Davis-County.html.

Munich RE (2013). "Floods dominate natural catastrophe statistics in the firsthalf of 2013." Retrieved from https://www.munichre.com/us/propertycasualty/press-news/press-releases/2013/130709-natcatstats-first-half-2013/ index.html.

Nath, J. (2012). "CLEAN SAFE WATER IS THE TRUE STAFF OF LIFE." Retrieved from https://www.watertechonline.com/home/article/15539750/clean-safe-water-is- the-true-staff-of-life

National Aeronautics and Space Administration (NASA) –EARTH OBSERVATORY. (2010). "Violent Storm Strikes Western Europe." Retrieved from http://earthobservatory.nasa.gov/NaturalHazards/view. php?id=42881.

National Aeronautics and Space Administration (NASA). "GLOBAL CLIMATE CHANGE: Vital Signs of the Planet." Retrieved from http:// climate.nasa.gov/scientific-consensus.

National Aeronautics and Space Administration (NASA). "A Look Back at a Decade of Fires." Retrieved from http://www.nasa.gov/mission_pages/ fires/main/modis-10.html.

National Academy of Engineering. (2007). "NAE Grand Challenges: 21st Century Innovations." Retrieved from http://www.engineeringchallenges. org/14373/GrandChallengesBlog/8275.aspx.

National Aids Trust (2015). "HIV IN THE UK STATISTICS – 2015." Retrieved from Http://www.nat.org.uk/we-inform/HIV-statistics/ UK-statistics.

National Centers for Environmental Information (NOAA). "Climate Change and Extreme Snow in the U.S." Retrieved from, https://www.ncdc.noaa.gov/news/climate-change-and-extreme-snow-us

National Oceanic and Atmospheric Administration's (NOAA) (2014).
National Wildlife Federation. (2017). "Global Warming: Threat to Wildlife." Retrieved from http://www.nwf.org/Wildlife/Threats-toWildlife/Global- Warming.aspx.

North Carolina State University. "Climate Education for K-12 - Global Warming vs. Climate Change." Retrieved from http://climate.ncsu.edu/ edu/k12/.gwvcc.

O'Neill, L. (2009) reliefweb. "Seeking survivors after Asian-Pacific catastrophes." Retrieved from, http://reliefweb.int/report/american-samoa/seeking-survivors-after-asian-pacific-catastrophes.

Passoth, K. (2015). "Birds mysteriously dying in El Reno." Retrieved from (http://www.koco.com/crime/Birds-mysteriously-dying-in-El-Reno/31019254).

Perry, H. (2015). "Chinese village awakens to a mystery of 100 tons of dead fish floating in their local pond." Retrieved from http://www.dailymail.co.uk/news/peoplesdaily/article-3037028/Something-fishy-s-goingChinese- village-awakes-mystery-100-tonnes-dead-fish-floating-local-pond. html.

Podesta, J. (2019). "The Climate Crisis, migration, and refugees." Retrieved from, https://www.brookings.edu/research/the-climate-crisis-migration-and-refugees/

Prada, P. & D. Kinch. (2011). The Wall Street Journal. "Brazil Landslide and Flood Toll." Retrieved from http://www.wsj.com/articles/SB1000142 4052748704029704576087893783940786.

PR Newswire (2017). "Key climate-change vulnerabilities identified for three St. John River communities." Retrieved from http://finance. yahoo.com/news/key-climate-change-vulnerabilities-identified-

Public Health England (2015). "HIV statistics talk about 'men who have sex' with men to include all gay, bisexual and other men who have sex with men." Retrieved from http://www.nat.org.uk/we-inform/HIV-statistics/ UK-statistics.

Ranson, M. (2014). "Crime, weather, and climate change." Journal of Environmental Economics and Management, 67: 274–302. Feb. 2014. Accessed on 07 April 2014.

Reagan, D. R. (2014). "The Gate to Prophecy." Retrieved from Rapture Ready. http://www.raptureready.com/.

Roach, J. (2009) NATIONAL GEOGRAPHIC NEWS. "Arctic Largely Ice-Free in Summer Within Ten Years?" Retrieved from Rogers, P.

(2014). Updated 2016).

Roberts, J. (2010). "Media Use Crazy Weather to Hype Global Warming, Despite Admissions Weather Isn't Climate." Retrieved from http://www.newsbusters.org/blogs/nb/julia-seymour/2010/08/19/ media-use- crazy-weather-hype-global-warming-despite-admissions.

Rosenzweig, C., D. Karolly, M. Vicarelli, P. Neofotis, Q. Wu, G. Casassa, A. Menzel, T. L. Root, N. Estrella, B. Seguin, P. Tryjanowski, C. Liu, S. Rawlins, & A. Imeson. (2007). "Attributing physical and biological impacts to anthropogenic climate change." Retrieved from Journal home, Advance online publication, http://www.nature.com/nature/journal/v453/ n7193/full/nature06937.html.

Sahney, S., Benton, M. J., & Falcon-Lang, H. J. (2010). "Rainforest collapse triggered Pennsylvanian tetrapod diversification in Euramerica." Geology. 38 (12

Santa Maria, C. (2014). "Study suggests tornado strength, frequency is increasing." Retrieved from News – EDITOR'S CHOICE, https://www.theweathernetwork.com/news/articles/study-links-tornado-strength-frequency-to-climate-change/33473/.

Santora, M. (2016). "Woman Thrown in Front of Train at Times Square Subway Station." The New York Times. Retrieved from https://www.nytimes.com/2016/11/08/nyregion/person-thrown-in-front-of-subwaytrain-is- killed-police-say.html?_r=1.

Schwartz, R. M. & T. W. Schmidlin. (2002). The American Meteorological Society. Journal Online. "Climatology of Blizzards in the Conterminous United States, 1959-2000." Retrieved from http://journals.ametsoc.org/doi/full/10.1175/1520-0442(2002)015%3C1765%3ACOBITC%3E2.0.C O%3B2.

Seftor, C. & L. Betz. NASA - Goddard Space Flight Center (Greenbelt, Maryland). Retrieved from http://www.nasa.gov/mission_pages/fires/ main/siberia-smoke.html.

Shaffer, G., S. M. Olsoe, J. O. P. Pedersen. (2009). Long-term depletion in response to carbon dioxide emissions from fossil fuels. Nature Geoscience 2, 105 - 109 (2009). Retrieved from (http://adsabs.harvard.edu/abs/2009NatGe … 2 … 105S).

Shambaugh, J. C. (2012). "The Euro Three Gates – Brooking Papers on Economic Activity, Spring 2012)" The Brooking Institution, Washington, DC 20036.

Simon, S. (2010). "EARTHQUAKES! What is going on?" Retrieved from http://www.seymoursimon.com/index.php/blog/ what_do_you_think_fact_or_fiction_are_bats_blind/P780/.

Sinha, V. (2010). "Refugee Camps Spread Life-Threatening Diseases." The *ReliefWeb Retrieved* from http://reliefweb.int/report/world/ refugee-camps- spread-life-threatening-diseases.

Smith, M. (2012). "Mass death of birds and fish – solar weapons test?" Retrieved from http://abundantmichael.com/blog/index.cfm/2011/1/5/ mass-death-of-birds- and-fish--scalar-weapons-test.

Snyder, M. (2015). "Will the Discovery of Huge Amounts of Oil in Israel Lead to War in the Middle East? "INFOWARS. Retrieved from https://www.infowars.com/will-the-discovery-of-huge-amounts-of-oil-in-israellead-to- war-in-the-middle-east/.

Snyder, M. (2013). "Why Are Millions of Fish Suddenly Dying in Mass Death Events All Over the Planet?" Retrieved from http://www.activistpost.com/2013/08/why-are-millions-of-fish-suddenly-dying.html.

Snyder, M. (2011). "End of The American Dream: The American Dream Is Becoming a Nightmare and Life as We Know It Is About to Change - hurricane-irene-unusual-earthquakes-unprecedented-tornadoes-

Sobelman, B. (2015). "One country that won't be taking Syrian refugees: Israel. "The Los Angeles Times". Retrieved from http://www.latimes.com/world/middleeast/la-fg-syrian-refugees-israel-20150906-story.html

Solomon, S., Plattner, G.K., Knutti, R., and Friedlingstein, P. (2008). "Irreversible climate change due to carbon dioxide emissions." Retrieved from http://www.pnas.org/content/106/6/1704.full.pdf.

Spaner, J. S.; LeBali, H. (2013). "The Next Security Frontier." Proceedings of the US Naval Institute. 139 (10): 30-35. Retrieved November 2015. https://en.wikipedia.org/wiki/Effects_of_global_warming#cite_note-131.

Spence, M. A. (2013). "What's Stopping Robust Recovery?" Retrieved from http://www.cfr.org/financial-crises/s-stopping-robust-recovery/p32080.

Spence, M. & D. Leipziger. (2013). "Globalization and Growth Implications for a PostCrisis World." Retrieved from https://www.scribd.com/document/30504860/Globalization-and-Growth-Implications-for-a-Post-Crisis-World.

Stevens, J. (2011). The Daily Mail. "Wiped off the map: Shocking before and after images reveal how giant tornado ripped apart Joplin's city landmarks." Retrieved from http://www.dailymail.co.uk/news/article-1389737/JoplinMO-tornado-At-89-dead-twister-cuts-4-mile-swathe-Missouri-town. html.

Stocker, et al - Intergovernmental Panel on Climate Change [ipcc], 2013). "Summary for Policymakers." Retrieved from http://www. climatechange2013.org/images/report/WG1AR5_SPM_FINAL.pdf.

Stormtrack (2007). "Tornadoes around the world. Want to know who gets the most tornadoes?" Retrieved from https://stormtrack.org/community/threads/tornadoes-around-the-world-want-to- know-who-gets-the-mosttornadoes.10669/.

Teslik, L. H. (2007). "Economic Problems at Every Turn." Retrieved from http://www.cfr.org/financial-crises/economic-problems-every-turn/p14746.

The Center for Disease Control and Prevention (CDC) (2021). "CORONAVIRUS – COVID-19." Retrieved from https://www.cdc.gov/coronavirus/2019-ncov/covid-data/forecasting-us.html

The *Economist.* (2013). "The march of protest around the world: a wave of anger is sweeping the cities of the world. Politicians beware." Retrieved from http://www.economist.com/news/leaders/21580143-waveangersweeping-cities- world-politicians-beware-march-protest.

The *Economist* (2012). "They did not have to be so unfair." Neil Barofsky's Bailout and Sheila Bair's Bull by the Horns. http://www.economist.com/blogs/freeexchange/2012/10/americas-bank-bailouts.

The *Extinction Protocol* (2015). "Are earthquakes and volcanic eruptions increasing across the planet?" Retrieved from https://theextinctionprotocol.wordpress.com/2015/04/28/are-earthquakes-and-volcanic- eruptions-increasing-across-the-planet/.

The *Indian Express* (2014). "Odisha floods kill 34, affect nearly 10 lakh people." Retrieved from http://indianexpress.com/article/india/india-others/odisha-floods- kill-34-affect-nearly-10-lakh-people/.

The Miami Herald. (2019). "The Bahamas have a tough road ahead. Dorian caused $3.4 billion worth of damage; report says." Retrieved from, https://www.miamiherald.com/news/nationworld/world/americas/article2374358 14.html

The Mud Report. (2013). "Climate Scientists All Know that 'Climate Trains the Boxer but Weather Throws the Punches." Retrieved from http://themudreport.blogspot.com/2013/07/climate-scientists-all-know-that. html.

The National Hurricane Center (NHC). Retrieved from (http://www.livescience.com/52575-hurricanepatricia-storm-strength.html).

The National Geographic Magazine (2012). "Weather gone wild." Retrieved from http://www.nationalgeographic.com/magazine/2012/09/extreme-weather-global- climate-change-effects/.

The *New York Times* (2016). "Piracy at Sea." Retrieved from https://www. nytimes.com/topic/subject/piracy-at-sea.

The *Sentinel* (2012). "Mystery surrounds the death of 5,000 fish." Retrieved from http://www.stokesentinel.co.uk/mystery-surrounds-death-

5-000fish/story- 16292504-detail/story.html#ixzz3ql456jQT.

The Straits Times (2021). "US intelligence says Russia likely behind hacking of government agencies, United States." Retrieved from US intelligence says Russia likely behind hacking of government agencies, United States News & Top Stories - The Straits Times

The United States Environmental Protection Agency, EPA). "Climate Change Impacts: Introduction to Global Issues." Retrieved from https://www.epa. gov/climate-impacts/international-climate-impacts.

The United Nations (2018). "Refugees." Retrieved from, http://www.unfoundation.org/search/search. jsp?query=Refugees&x=0&y=0

The United Nations. (2016). "Global Issues Overview." Retrieved from http://www.un.org/en/sections/issues-depth/global-issues-overview/

The United Nations. "MARITIME PIRACY." Retrieved from http:// www.unodc.org/documents/data-and-analysis/tocta/9.Maritime_piracy. pdf.

The United Nations. (2016). "Refugee crisis: Europe and the Middle East." Retrieved from https://www.rescue.org/topic/refugee-crisis-europe-middle-

The Wall Street Journal. (2009). "The Great Recession and Government Failure." Retrieved from http://www.wsj.com/articles/SB10

The Washington Post. (2015). "Passengers watched killing on Metro car. Should they have intervened?" The Washington Post. Retrieved from https://www.washingtonpost.com/local/witnesses-to-a-stabbing-didnt-confront- killerdo-they-deserve-condemnation/2015/07/08/

Modern Piracy – History of Piracy - The Way of Pirates (2016). "Way Pirates". Retrieved from www.thewayofthepirates.com/piracy-history/modern-piracy/.

Tignino, M. (2010). "Water, international peace, and security." INTERNATIONAL REVIEW OF THE Red Cross. Volume 92, Number 879, September 2010. Retrieved from www.icrc.org /eng/assets/files/

Tippett, M. (23016). "Extreme Tornado Outbreaks Have Become More Common." Retrieved from International Research Institute for Climate and Society, http://iri.columbia.edu/news/tornado-outbreaks/.

Tooley, H. (2015). INQUISITR NEWS. "1,300 DEAD BIRDS FOUND ON BEACH: MYSTERIOUS DEATH OF SEABIRDS IN CHILE NARROWED DOWN." Retrieved from http://www.inquisitr.com/2101228/1300-dead-birds-found-on-beach-mysterious-death-of-seabirds-in-chile-narrowed-down/.

Trenberth, K. (2007). "HURRICANES, TYPHOONS, CYCLONES: Background on the science, people, and issues involved in hurricane research." Retrieved from NCAR, https://www2.ucar.edu/news/backgrounders/hurricanes-typhoons-cyclones#8.

Tsosie, R. (2007). "Indigenous People and Environmental Justice: The Impact of Climate Change." University of Colorado Law Review. 78:1625. Retrieved from Wikipedia, #cite_note-ref_10-6

Tyson, P. (2003). "Would a dramatic change in the Earth's magnetic field affect creatures that rely on it during migration?" Retrieved from http://www.pbs.org/wgbh/nova/nature/magnetic-impact-on-animals.html.

University Corporation for Atmospheric Research (UCAR) (2017). Retrieved from https://www2.ucar.edu/.

UCS: Science for a Safer World. "Global Temperatures and Climate Change." Retrieved from http://www.ucsusa.org.

Union of Concerned Scientists, 2012). "Flooding events are becoming more common in a warming world." Retrieved from http://www.ucsusa.org/global_warming/science_and_impacts/impacts/global-warming- andflooding.html#.WKylpPWcGUk.

United States Geological Survey (USGS). "2010 Significant Earthquakes and Main Headlines Archive." Retrieved from www.earthquake.usgs.gov.

USAID (2013). "Typhoon Haiyan (Yolanda)." Retrieved from https:// www.usaid.gov/haiyan. U.S. Government document 12039-Historical

Tables). Retrieved from www.cbo.gov/ftpdocs/12oxx/doc12039/HistoricalTables [1].pdf.

USGS. "Earthquakes." Retrieved from http://earthquake.usgs.gov/ earthquakes/.

VICE News, Vietnam (2016). "Mysterious Mass Fish Death Sparks Rare Public Protestin Repressive Vietnam." Retrieved from https://news.vice.com/article/vietnam-breaks-up-protests-shuts-down-social-media-amidpublic- anger-over-massive-fish-kill.

Vidal, J. (2014). World begins 2014 with unusual number of extreme weather events. The Guardian. Retrieved from https://www.theguardian. com/environment/2014/feb/25/world-2014-extreme-weather-events.

Warrick, J. (2016). "Mysterious mass deaths of Alaskan birds' baffles Scientists." The Washington Post. Retrieved from https://www.washingtonpost.com/news/energy-environment/wp/2016/01/12/mysterious-mass-deaths-of-alaskanbirds-baffles- scientists/?utm_term=.77fcf4305d4b.

WebMD (2020). "COVID-19 Vaccines." Retrieved from https://www.webmd.com/vaccines/covid-19-vaccine/news/20201214/closer-look-at-three-covid-19-vaccines

Welch, C., (2015). "Mass Death of Seabirds in Western U.S. Is 'Unprecedented'" National Geographic. Retrieved from http://news.nationalgeographic.com/news/2015/01/150123-seabirds-mass-die-offauklet- california-animals-environment/.

Werrell, C. & Femia, F. (2015). "Climate Change in the UK National Security Strategy & Strategic Defence & Security Review 2015." Retrieved from, https://climateandsecurity.org/2015/11/24/climate-change-in-the-uk-national- security-strategy-strategic-defence-security-review-2015/

West, N. and Z. Gardner. (2011). The Activist Post. "The 10 Leading Theories for Dead Birds and Fish." Retrieved from http://www.activistpost.com/2011/01/10-leading-theories-for-dead-birds-and.html.

Wikipedia, the free encyclopedia. Information used retrieved from, https://en.wikipedia.org/

Wikipedia, the free encyclopedia (2021). "Storming of the United States Capitol." Retrieved from https://en.wikipedia.org/wiki/2021_storming_of_the_United_States_Capitol#:~:text=Part%20of%20wider%20protests%2C%20it,the%20Capitol%2C%20and%20five%20deaths.

Wildenboer, N. (2013). "Test to be conducted on Dead Birds." Retrieved- http://www.iol.co.za/scitech/science/environment/tests-to-beconducted-on-dead- birds-1.1617937#.VdnwDU3H_cs.

Williams, A. (2014). "Experts baffled by the cause of bird deaths." Herald Live. Retrieved from http://www.heraldlive.co.za/ experts-baffled-by-cause-of-bird- deaths/.

Wing, N. (2015). "Guns kill an Average of 36 People Every Day, And the Nation Doesn't Even Blink." Retrieved from http://www.huffingtonpost. com/entry/us- gun-violence_us_560d9635e4b0af3706e00e20.

Wood, A. (2013). "Alarm as thousands of dead birds washed up on the east coast." The Yorkshire Post. Retrieved from http:// www.yorkshirepost.co.uk/news/mtain- topics/general-news/alarm-as-thousands-of-dead-birds-washed-up-on-east-coast- 1-5564156.

World Economic Forum, 2019). "Future of Economic Progress." Retrieved from - https://www.weforum.org/agenda/archive/economic-growth-and-social- inclusion

WORLD ECONOMIC FORUM. "Global Leadership Fellows." Retrieved from, https://www.weforum.org/communities/global-leadership-fellows.

World Health Organization (WHO) "Coronavirus disease (COVID-19)." Retrieved from https://www.who.int/health-topics/coronavirus#tab=tab_1

World Health Organization (WHO) (2019). "Infectious Diseases." Retrieved from, https://www.bing.com/search?q=WHO+infectious+diseases

World Health Organization (WHO) (2014). "The EBOLA Virus." Retrieved from http://www.bbc.com/news/world-africa-29615452.

World Leadership School (2019). "Collaborative Leadership Programs." Retrieved from, http://worldleadershipschool.com/student-travel

WORLD METEOROLOGICAL ORGANIZATION. "Climate Change." Retrieved from https://public.wmo.int/en/our-mandate/climate.

World News (2020). "Turkey's earthquake kills 22 people; the interior minister says." Retrieved from, https://www.upi.com/Top_News/World-News/2020/01/25/Turkeys-earthquake-kills-22-people-interior-minister- says/5201579961116/

Wright, P. (2016). "Matthew-Fueled Storm System Causes Severe Flooding in Eastern Canada." Retrieved from https://weather.com/news/news/matthew-canada-flooding-newfoundland-nova-scotia.

Yohe, G. and C. Parmesan. (2003). "A globally coherent fingerprint of climate change impacts across natural systems." Retrieved from http://gyohe.faculty.wesleyan.edu/files/2010/11/Parmesan-Yohe-Nature.pdf.

Zakaria, F. (2016). CNN Sunday Morning presentation of GPS.

About the Author:

Desmond Michael Coverley was born in Grand Turk, Turks and Caicos Islands, and is commonly known as *Brother Des.*

Brother Des' educational pursuits include a high school diploma from H. J. Robinson High School, Grand Turk; a degree in *Religious and Christian Education* from Bluewater Bible College & Institute, St Thomas USVI (1972); Associate in Applied Science degree in *Radiologic Technology* from the University of the District of Columbia (UDC), Washington DC (1983); a BA degree in *Physician Assistantship (PA)* from Howard University, Washington DC (1987); a Master's degree in *Clinical & Community Psychology* from the University of the District of Columbia (UDC), Washington DC (1993); and a Ph.D. in *Organizational Communication & Health Communication* from Howard University, Washington DC (2008). He also received training and certifications in *Substance Abuse* at the University of the West Indies, Trinidad (1997); *Iridology* in California (2002-2003 & 2007); and *Distance Learning* at Howard University, Washington, DC (2013).

Brother Des lectured at Howard University as an Associate Professor in sciences for 18 years, at which time he served the National Society of Allied Health (NSAH) as President-Elect (2007-2009) and President (2009-2012). He is presently serving as a Member of the Board.

Brother Des has engaged in Christian Education. He was ordained to the Gospel Ministry (1974); Pastor of the Bible Baptist Church (8 Yrs.); Founded the Grand Turk Christian Academy, Grand Turk, TCI.; Founded and pastored Bible Study Ministries for many years – TCI & Maryland. Also, while performing his ministries, he founded the Drug Unit and Directed the National Drug Council of the Turks and Caicos Islands. He holds an Indefinite License in Christian Ministries from The Supreme Court of the District of Columbia, Washington DC since 1991.

Brother Des is a Prophetic Bible Teacher and Author. He has published several books: (Hard Covers, Paperbacks, & eBooks) *(1) Global Warming Or God's Warning (Part 1); (2) Global Warming Or God's Warning (Part 2); (3) The Mystery: Mortals Becoming Immortal; (4) The Coming Great Tribulation On The Earth; (5) Armageddon; (6) The Only True And Righteous Judge; (7) The Last Leader is Coming; (8) The Bible Explains the Climate Change Controversy; (9) Climate Change and Signs of the End Time; (10) The Archival History of the National Society of Allied Health; (11) Climate Crisis: A Prophetic Explanation.* Books may be purchased online at Amazon.com; Barnes and Noble; and WestBow Press. **His Slogan is, Read and Be Ready!**

www.ingramcontent.com/pod-product-compliance
Lightning Source LLC
Chambersburg PA
CBHW061503180526
45171CB00001B/16